Soil Science: Nature and Properties of Soil

Soil Science: Nature and Properties of Soil

Leyton Gray

SYRAWOOD
PUBLISHING HOUSE

New York

Published by Syrawood Publishing House,
750 Third Avenue, 9th Floor,
New York, NY 10017, USA
www.syrawoodpublishinghouse.com

Soil Science: Nature and Properties of Soil
Leyton Gray

International Standard Book Number: 978-1-68286-821-8 (Hardback)

Cataloging-in-Publication Data

Soil science : nature and properties of soil / Leyton Gray.
   p. cm.
Includes bibliographical references and index.
ISBN 978-1-68286-821-8
1. Soil science. 2. Soils. I. Gray, Leyton.
S591 .S65 2019
631.4--dc23

TABLE OF CONTENTS

PREFACE

Soil is a composition of organic matter, liquids, gases and minerals that together is conducive to the subsistence of life. Some of the common minerals in soil are calcite, quartz, mica, feldspar, etc. The role of soil is significant as a medium that supports and promotes plant growth and development, stores water and also acts as a habitat for organisms. It also acts as a modifier of the Earth's atmosphere. The soil undergoes a constant process of development in terms of various physical, biological and chemical processes, such as weathering and erosion. The study of the soil is approached from the two branches of pedology and edaphology. Edaphology investigates the influence of soil on living organisms, while pedology is concerned with the formation, classification and morphology of the soil. The properties of the soil, which are crucial for ecosystem services, include bulk density, texture, porosity, structure, temperature, consistency, etc. This textbook provides comprehensive insights into the study of soil. It is a compilation of chapters that discuss the most vital concepts about the nature and properties of soil. It is an essential guide for both academicians and students who wish to pursue this discipline further.

A detailed account of the significant topics covered in this book is provided below:

Chapter 1- The soil is a mixture if minerals, organic matter, liquids, gases and organisms that support life. It supports plant growth and stores, purifies and supplies water. The aim of this chapter is to provide an introduction to soil. It includes topics like soil, moisture condition, soil pH and soil fertility which are crucial for a complete understanding of the soil.

Chapter 2- In order to understand the suitability of a particular soil or to determine the relationship between varied soils, the soil is classified into different categories. The topics elaborated in this chapter include clay soil, sandy soil, red soil, silt soil, topsoil, alkali soil, etc. which will help in developing a better perspective about the classification of soil.

Chapter 3- Pedology is the science that is concerned with the classification, description and formation of soils. It is a branch of soil science. This chapter discusses in detail the significant aspects and principles of pedology. It includes vital topics such as pedogenesis, soil morphology, active layer and soil map.

Chapter 4- Edaphology is a branch of soil science concerned with the study of the influence of soils on plants and other living organisms. This chapter has been carefully written to provide an easy understanding of the varied facets of edaphology, such as agricultural soil science, environmental soil science, bioeffector, plant nutrition, phosphate solubilizing bacteria, etc. for a holistic understanding of edaphology.

Chapter 5- The properties of the soil such as soil texture, structure, resistivity, bulk density, etc. play a crucial role in ecosystem services. The study of the soil requires an understanding of its varied properties, such as soil structure, soil resistivity, soil texture, pore space of soil, thermal properties, etc. which have been carefully elaborated in the chapter.

Chapter 6- The soil stores and controls the release of nutrients. This is referred to as the soil nutrient cycle. In order to completely understand the soil nutrient cycle, it is necessary to understand the processes related to it, such as ion absorption by plant roots, carbon cycle in soil, nitrogen cycle in soil, etc. which have been dealt in extensive detail in this chapter.

Chapter 7- The measure of the soil condition relative to its requirement, whether to sustain plant or animal life, support human health or to enhance air and water quality is termed as soil quality. This chapter has been written in order to help develop a better understanding of soil quality. It includes the topics related to this area, such as soil tests for determining soil quality, soil respiration and bulk density of the soil.

Chapter 8- The organic matter content of the soil that consists of animal and plant residues at all their stages of decomposition, soil organisms along with their cells and tissues together is known as soil organic matter. The topics covered in this chapter such as, soil organic matter component and management of soil organic matter are crucial for a deeper understanding of soil.

Chapter 9- The displacement of the topmost layer of the soil is termed as soil erosion. It is caused due to the activity of water, glaciers, wind, living organisms, etc. The effect of soil erosion on vegetation and slope stability as well as the environment and the ways in which erosion can be managed have been dealt with in this chapter.

Chapter 10- The contamination of the soil caused due to any change to the natural soil environment or by the presence of xenobiotic chemicals in the soil is termed as soil pollution. It is caused due to an improper disposal of waste, industrial activities or agricultural chemicals. This chapter analyzes the causes of soil pollution, its effects and the measures to control soil pollution.

It gives me an immense pleasure to thank our entire team for their efforts. Finally in the end, I would like to thank my family and colleagues who have been a great source of inspiration and support.

Leyton Gray

Chapter 1

Introduction to Soil

The soil is a mixture if minerals, organic matter, liquids, gases and organisms that support life. It supports plant growth and stores, purifies and supplies water. The aim of this chapter is to provide an introduction to soil. It includes topics like soil, moisture condition, soil pH and soil fertility which are crucial for a complete understanding of the soil.

Soil is normally considered as the fine earth which covers land surfaces as a result of the in situ weathering of rock materials or the accumulation of mineral matter transported by water, wind, or ice. The distinctive feature of soil is that to this weathered mineral material is added organic material. This organic material may be both living and dead. The dead organic matter will include little altered and freshly added dead plant roots and leaf and other plant litter, dead fauna, and organic material in various stages of decomposition from little modified relatively fresh materials to the complex decomposed material called humus. It is this mixture of mineral and organic material which gives the soils their distinctive characteristics. Across the surface of the earth there are many different types of soil which reflect, at least in part, varying combinations of mineral and organic matter and their differing responses–both separately and often in complex association–to different environmental conditions. Indeed soil (and the soil constituents), together with the plant life it supports, the rock on which it lies, and the climate it experiences, forms a finely balanced system.

Depending upon the context, the word "soil" may have very different meanings. A simple definition of soil is the material that plants grow in and which provides them with physical support, water, and nutrients. There are other more particular uses of the term soil. To the engineer, soil is the finely divided and relatively loose "rock" material at the earth's surface, and this over burden is often considered an inconvenience because it must be removed. The geologist calls this layer there goliath and geological investigations frequently begin below it. The hydrologist looks on the soil as if it were a large reservoir storing water to supply streams and rivers. The most widely held view of the soil, however, is as a medium for plant growth and the provision of food and fiber directly or through the intermediate stage of animals.

Soil is important as a medium for plant growth and for the support of much animal and human activity. The soil acts as a reservoir for nutrients and water providing the plants' needs for these requirements throughout their growth. The soil may also provide an environment for the break- down and immobilization of materials added to the surface (in addition to the aforementioned plant and animal remains) such as fertilizers and pesticides and waste products such as sewage sludge, animal wastes and slurries, and composted refuse materials. The soil is a complex dynamic system in which the interactions of the biological, chemical, and physical environments results in the transformation of materials, possibly rendering initially harmful materials less dangerous and immobilizing others as a result of the interactions between these added materials and the organic and inorganic soil constituents. This immobilization may enable breakdown of the potentially dangerous materials to less dangerous forms. These interactions and transformations may be long term, over decades, medium term over months or years, short term between individual events such as rainstorms, or almost instantaneous.

The soil usually consists of a vertical sequence of layers or horizons differentiated by physical, chemical, or biological characteristics. This sequence of layers is normally described as the soil profile. Surface horizons are usually characterized by accumulations of organic materials as a result of the addition of plant and animal residues (it is, however, becoming increasingly apparent that organic additions in deeper layers from roots may have a significant role to play in the overall additions of organic material and in the operation of the soil processes). Surface horizons with a mixture of mineral and organic materials are usually identified as A horizons and may be further described by the amount and nature of the organic matter present within the horizon. Where there is a substantial proportion of organic material in the surface layer it may be called O horizon. The A and O horizons are generally considered the biologically most active part of the soil and are also subject to considerable changes as a result of the weather, being frequently exposed to wetting and drying cycles and to a wide range of temperatures, which results in high rates of activity in physical and chemical processes. Both organic and mineral materials are transformed in these surface layers and frequently some of the byproducts of these trans-formations are released and removed to lower layers within the soil (eluviation). Below the A horizons usually B horizons are encountered where the balance shifts from predominantly biological processes as in the A horizons to predominantly chemical and physical processes. Pedologists normally consider this zone to be that of peak activity of the Pedological or soil-forming processes, and it is for this reason that many of the currently used soil classifications pay particular attention to the characteristics of the B horizon in the allocation of soils to classes. The B horizons are frequently subject to the maximum intensity of weathering processes and trans-formations and may also exhibit the accumulation of materials trans located from surface layers (illuviation) such as clays, organic matter, and iron and aluminum hydrous oxides. The description of B horizons frequently emphasizes the degree of weathering and the nature of the accumulated materials. The C horizon, below the B, is the relatively unaltered material from which the A and B horizons may have developed.

The importance of the incorporation and breakdown of organic matter and the production of complex organic materials, as well as the weathering of mineral materials to produce secondary products (such as clay minerals) and to release minerals which may be taken up by plants or leached out of the system, provides the distinctive nature of the soil. In particular the large surface area associated with soil materials and their ability both to retain and exchange ions (both those added in fertilizers and those released during weathering and biological decomposition) and to absorb water are important properties which distinguish soil from rock materials.

The soil performs many functions. These include functions related to natural ecosystems, agricultural productivity, and environmental quality, soil as source of raw materials and as base for buildings. Of these the agricultural productivity function is probably the most widely recognized and understood. This function of soil is to support plant and animal productivity whilst maintaining or enhancing water and air quality and also supporting human health and habitation. To per-form this production function, the soil must be able to provide the following: a physical, chemical and biological context suitable for the survival and development of living organisms;

the conditions for the regulation and partitioning of water flow, storage and recycling of nutrients and other elements; conditions to support biological activity and diversity for plant growth and animal productivity; an environment to filter, buffer, degrade, immobilize, and detoxify organic and inorganic substances; and provide mechanical support for living organisms and their structures. This brief list of requirements illustrates the complex demands placed upon soil if it is to maintain the productivity which is essential if the food demands of an ever growing global population are to be met. It further illustrates the need to understand the complex interactions between the various components within the soil and the interactions with other components of the terrestrial ecosystems. The productivity functions of soil is probably the most widely studied and understood, but the soil is increasingly expected to perform functions related to the maintenance of environmental quality, such as the filtering of toxic materials applied to the soil surface. If the soil is to perform these varied functions effectively and with no reduction in its overall quality the role of the soil, its components and processes must be investigated and understood. It must be ensured that in performing one function the soil's ability to perform other functions is not reduced or removed.

Soil is essential for many human activities. It is also a basic part of the natural environment. The development of humans and society since prehistoric times has been closely linked with an increasing ability to manage the soil to human benefit. This progress has been achieved by adjusting the balance between the soil and its natural environment. These adjustments have not always resulted in positive responses and benefits, and there are records throughout history describing soil destruction as a result of mismanagement and misuse of the soil, often as a result of the failure to understand the nature and complexity of the soil – environment relationships.

Soil Moisture Condition

The soil moisture content indicates the amount of water present in the soil.

It is commonly expressed as the amount of water (in mm of water depth) present in a depth of one meter of soil. For example: when an amount of water (in mm of water depth) of 150 mm is present in a depth of one meter of soil, the soil moisture content is 150 mm/m.

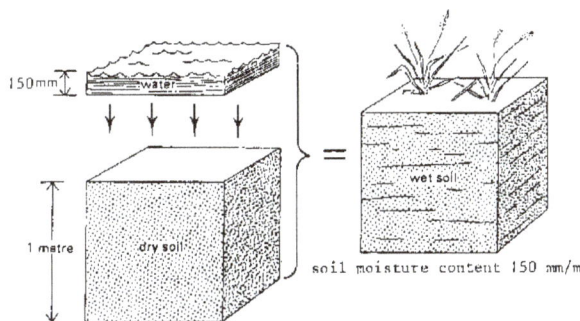

Fig. A soil moisture content of 150 mm/m.

The soil moisture content can also be expressed in percent of volume. In the example above, 1 m3 of soil (e.g. with a depth of 1 m, and a surface area of 1 m²) contains 0.150 m3 of water (e.g.

with a depth of 150 mm = 0.150 m and a surface area of 1 m2). This results in soil moisture content in volume percent of:

$$\frac{0.150\,\mathrm{m^3}}{1\,\mathrm{m^3}} \times 100\% = 15\%$$

Thus, a moisture content of 100 mm/m corresponds to a moisture content of 10 volume percent.

Note: The amount of water stored in the soil is not constant with time, but may vary.

Saturation

During a rain shower or irrigation application, the soil pores will fill with water. If all soil pores are filled with water the soil is said to be saturated. There is no air left in the soil. It is easy to determine in the field if a soil is saturated. If a handful of saturated soil is squeezed, some (muddy) water will run between the fingers.

Plants need air and water in the soil. At saturation, no air is present and the plant will suffer. Many crops cannot withstand saturated soil conditions for a period of more than 2-5 days. Rice is one of the exceptions to this rule. The period of saturation of the topsoil usually does not last long. After the rain or the irrigation has stopped, part of the water present in the larger pores will move downward. This process is called drainage or percolation.

The water drained from the pores is replaced by air. In coarse textured (sandy) soils, drainage is completed within a period of a few hours. In fine textured (clayey) soils, drainage may take some (2-3) days.

Field Capacity

After the drainage has stopped, the large soil pores are filled with both air and water while the smaller pores are still full of water. At this stage, the soil is said to be at field capacity. At field capacity, the water and air contents of the soil are considered to be ideal for crop growth.

Permanent Wilting Point

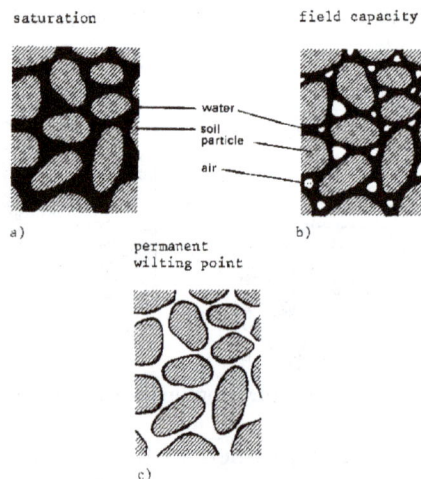

Fig. Some soil moisture characteristics

Little by little, the water stored in the soil is taken up by the plant roots or evaporated from the topsoil into the atmosphere. If no additional water is supplied to the soil, it gradually dries out.

The dryer the soil becomes, the more tightly the remaining water is retained and the more difficult it is for the plant roots to extract it. At a certain stage, the uptake of water is not sufficient to meet the plant's needs. The plant loses freshness and wilts; the leaves change color from green to yellow. Finally the plant dies.

The soil water content at the stage where the plant dies is called permanent wilting point. The soil still contains some water, but it is too difficult for the roots to suck it from the soil.

Available Water Content

The soil can be compared to a water reservoir for the plants. When the soil is saturated, the reservoir is full. However, some water drains rapidly below the root zone before the plant can use it.

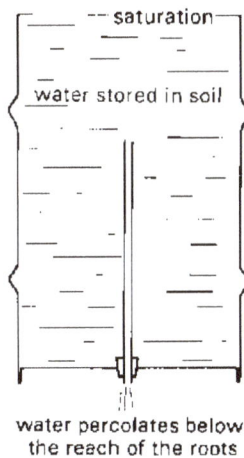

Fig. Saturation

When this water has drained away, the soil is at field capacity. The plant roots draw water from what remains in the reservoir.

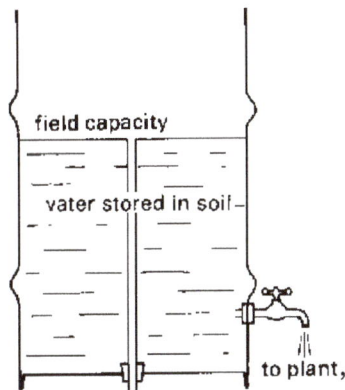

Fig. Field capacity

When the soil reaches permanent wilting point, the remaining water is no longer available to the plant.

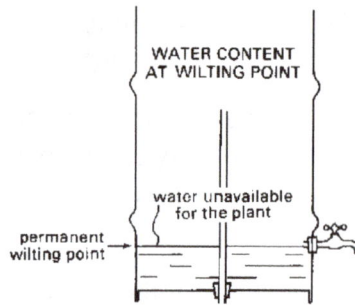

Fig. Permanent wilting point

The amount of water actually available to the plant is the amount of water stored in the soil at field capacity minus the water that will remain in the soil at permanent wilting point. This is illustrated in below.

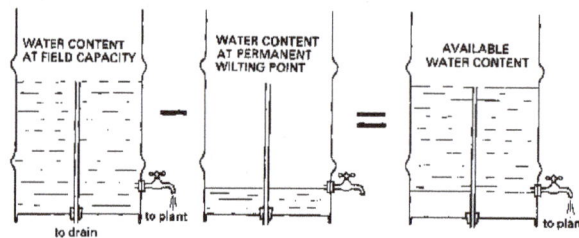

Fig. The available soil moisture or water content

Available water content = water content at field capacity - water content at permanent wilting point.

The available water content depends greatly on the soil texture and structure. A range of values for different types of soil is given in the following table.

Soil	Available water content in mm water depth per m soil depth (mm/m)
sand	25 to 100
loam	100 to 175
clay	175 to 250

The field capacity, permanent wilting point (PWP) and available water content are called the soil moisture characteristics. They are constant for a given soil, but vary widely from one type of soil to another.

Soil pH

Soil pH is a measure of the acidity or alkalinity in the soil. It is also called soil reaction.

The most common classes of soil pH are:

Extremely acid	3.5 – 4.4
Very strongly acid	4.5 – 5.0
Strongly acid	5.1 – 5.5

Moderately acid	5.6 – 6.0
Slightly acid	6.1 – 6.5
Neutral	6.6 – 7.3
Slightly alkaline	7.4 – 7.8
Moderately alkaline	7.9 – 8.4
Strongly alkaline	8.5 – 9.0

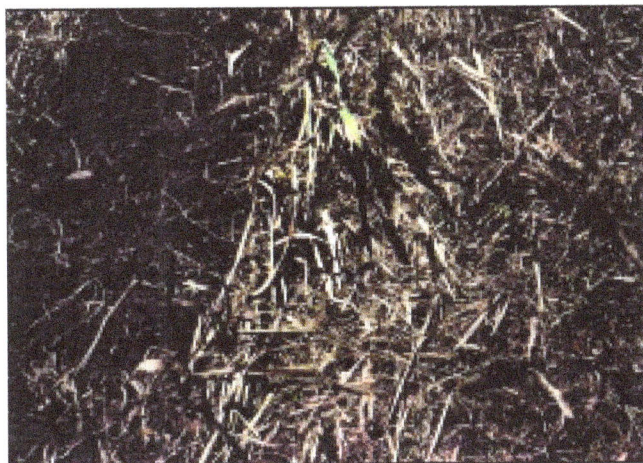

Significance of pH

Availability of Nutrients

Soil pH influences the solubility of nutrients. It also affects the activity of micro-organisms responsible for breaking down organic matter and most chemical transformations in the soil. Soil pH thus affects the availability of several plant nutrients.

A pH range of 6 to 7 is generally most favorable for plant growth because most plant nutrients are readily available in this range. However, some plants have soil pH requirements above or below this range.

Soils that have a pH below 5.5 generally have a low availability of calcium, magnesium, and phosphorus. At these low pH's, the solubility of aluminum, iron, and boron is high; and low for molybdenum.

At pH 7.8 or more, calcium and magnesium are abundant. Molybdenum is also available if it is present in the soil minerals. High pH soils may have an inadequate availability of iron, manganese, copper, zinc, and especially of phosphorus and boron.

Micro-organisms

Soil pH affects many micro-organisms. The type and population densities change with pH. A pH of 6.6 to 7.3 is favorable for microbial activities that contribute to the availability of nitrogen, sulfur, and phosphorus in soils.

Pesticide Interaction

Most pesticides are labeled for specific soil conditions. If soils have a pH outside the allowed range, the pesticides may become ineffective, changed to an undesirable form, or may not degrade as expected, which results in problems for the next crop period.

Mobility of Heavy Metals

Many heavy metals become more water soluble under acid conditions and can move downward with water through the soil, and in some cases move to aquifers, surface streams, or lakes.

Corrosivity

Soil pH is one of several properties used as a general indicator of soil corrosivity. Generally, soils that are either highly alkaline or highly acid are likely to be corrosive to steel. Soils that have pH of 5.5 or lower are likely to be highly corrosive to concrete.

Sources that Controls Soil pH

The acidity or alkalinity in soils has several different sources. In natural systems, the pH is affected by the mineralogy, climate, and weathering. Management of soils often alters the natural pH because of acid-forming nitrogen fertilizers, or removal of bases (potassium, calcium, and magnesium). Soils that have sulfur-forming minerals can produce very acid soil conditions when they are exposed to air. These conditions often occur in tidal flats or near recent mining activity where the soil is drained.

The pH of a soil should always be tested before making management decisions that depend on the soil pH.

Measurement of pH

A variety of kits and devices are available to determine the pH in the field. The methods include:

- dyes
- paper strips
- glass electrodes.

Soil pH can change during the year. It depends on temperature and moisture conditions, and can vary to as much as a whole pH unit during the growing season. Since pH is a measure of the hydrogen ion activity [H+], many different chemical reactions can affect it. Temperature changes the chemical activity, so most measurements of pH include a temperature correction to a standard temperature of 25 degrees C (770 F). The soil pH generally is recorded as a range in values for the soil depth selected.

Modification of Soil pH

A soil pH below about 5.6 is considered low for most crops. Generally, the ideal pH range is between 6.0 and 7.0. Liming is a common method to increase the pH. It involves adding finely ground limestone to the soil. The reaction rate for limestone increases when soil temperatures are warm and soil moisture is high. If the limestone is more finely ground, the reaction is faster.

The amount of limestone to apply depends on the amount of organic matter and clay as well as the pH. Fertility testing laboratories that have local experience make this determination.

A soil pH that is more than about 8.0 is considered high for most crops. Soils that have a pH in this range are often also calcareous.

Calcareous soils have a high content of calcium carbonate. The pH of these soils does not change until most of the calcium carbonate is removed. Acids that are added to the soil dissolve the carbonates and lower the soil pH. Treatments with acid generally are uneconomical for soils that have a content of calcium carbonate of more than about 5%. Because phosphorus, iron, copper, and zinc are less available to plants in calcareous soils, nutrient deficiencies are often apparent. Applications of these nutrients are commonly more efficient than trying to lower the pH.

When the soil pH is above 8.6, sodium often is present. These soils generally do not have gypsum or calcium carbonates, at least not in the affected soil horizons. Addition of gypsum followed by leaching using irrigation is a common reclamation practice. However, salts flushed into drainage water may contaminate downstream waters and soils.

The application of anhydrous ammonia as a nitrogen fertilizer contributes to lowering the soil pH. In some parts of the country, applications of ammonia lower the surface soil pH from ranges of 6.6 to 7.3 to below 5.6. This reduction can be easily overlooked in areas of no-till cropping unless the pH is measured in the upper 2 inches.

Chemical amendments that contain sulfur generally form an acid, which lowers the soil pH.

Soil Fertility

Soil fertility may be defined as the ability of soil to provide all essential plant nutrients in available forms and in a suitable balance whereas soil productivity is the resultant of several factors such as soil fertility, good soil management practices availability of water supply and suitable climate.

The soil is a natural medium for plant growth and it supplies nutrients to plants. Some soils are productive and they support luxuriant growth of plants with very little human effort whereas others may be unproductive which support almost no useful plant life regardless of every human effort. In order for soil to be productive, it must:

(i)   Be easily tillable and fertile,

(ii) Contain all essential elements in the forms readily available to plants in sufficient amount, and

(iii) Physically good to support plants and contain just the right amount of water and air for proper root growth. The soil must supply these essentials every day in the life of the plant.

Soil fertility and soil productivity appear to be synonymous but in soil science these two terms bear different meanings. Soil fertility may be defined as the ability of soil to provide all essential plant nutrients in available forms and in a suitable balance whereas soil productivity is the resultant of several factors such as soil fertility, good soil management practices availability of water supply and suitable climate.

A soil can be highly fertile, i.e., it has ready supply of nutrients in available form, yet it may not be highly productive. Water-logged soils may be highly fertile but may not produce good crop because of the un-favorable physical conditions.

A fertile soil may be highly saline or alkaline which may not be good for agriculture. Sandy soil may be poor in fertility but with the use of fertilizers and water it may be made productive. Soil fertility thus denotes the status of plant nutrients in the soil whereas the soil productivity is the resultant of various factors influencing crop production. In fact there is no standard for either fertility or productivity because both depend upon the crops to be grown. Soil that is productive for potatoes may not necessarily be productive for certain other crops.

Soil Fertility Factors

Several factors are known to govern the fertility of soil. Some of the important factors are discussed below:

1.   As a result of cropping, a large amount of organic matter and soil minerals are removed and if the normal cycling of mineral elements is retarded, loss in soil fertility may result.

2.   Besides cropping, soil erosion and loss of water also causes tremendous loss of plant nutrients from the top soil.

     Generally, water is lost through leaching, drainage, evapotranspiration and runoff.

     The following adverse effects are observed due to water loss:

     (i)   Soil becomes very hard.

     (ii) The seed germination percentage decreases.

     (iii) The nutrients in the soil leach or evaporate.

(iv) The root growth retards, so that plants become stunted and, as a result, the yield is reduced.

(v) Stomata become closed, as a result of which accumulation of gases or metabolic wastes increases in plant tissues leading to death of the plant.

(vi) The activity of soil microorganisms decreases.

3. Conversion of organic forms of nitrogen locked in humus into ammonia gas and nitrogen gas and leaching out of soluble nitrates and nitrites from surface soil greatly affect the fertility status of soil.

4. Like deficiency, the abundance of certain nutrient elements in soluble form may also be toxic and even the elements, say alkalies, essential for plant growth may be toxic if present in excess. Flowering plants do not grow in the soil containing more than 6 per cent NaCl and other salts. The elements are not equally toxic and the various species of plants differ in their susceptibility to different elements.

5. Toxic chemicals and pesticides in soil: Several agricultural chemicals being used for controlling various diseases and insect pests are highly toxic and their application adversely affects the soil micro flora and fauna. Prolonged persistence of these pesticides in soil is bound to lower the soil fertility both directly and indirectly.

6. Soil reaction: The soils may be alkaline or neutral or acidic in their reaction. Some plants find acid soil unsuitable for growth and other plants find alkaline ground un-favorable. pH value of soil solution determines the availability of certain plant nutrients and thus it has bearing on soil fertility problem. Increase in the acidity of the soil makes mineral salts more soluble in soil solution and thus salts may become available in concentrations that may be highly toxic or may damage plants growing in such soils. Janick et al. (1969) have demonstrated that high concentration of both iron and aluminium may damage plants growing in acid soils.

Maintenance of Soil Fertility

Soil fertility is the most important asset of a nation. Maintenance of soil fertility is an important aspect of agriculture. The soil fertility problem has been studied in many countries and scientists have brought to light several facts concerning soil fertility and its maintenance.

Soil Fertility is of two Types

(a) Permanent Fertility

It is derived from the soil itself. It can be improved, maintained or corrected by soil management practices.

(b) Temporary Fertility

It is acquired by suitable soil management but the response of built up soil fertility is highly dependent on the degree of permanent fertility which is already there. Several methods are known for controlling the loss of soil fertility. Here only the important methods are discussed.

Application of Organic Manures and Chemical Fertilizers

Plants absorb water and minerals from the soil, which is essential for growth, flowering, crop yield, and other vital activities. Soil is a store house for organic and inorganic plant nutrients. Some soils are rich in organic and humus content and are considered to be fertile and more productive while others that are deficient in humus and minerals are less productive.

The soil is subjected to a continuous depletion of nutrients due to its continuous use by crops. This requires the addition of mineral resources. The various soil components are being removed by living organisms and are returned to the soil by death and decay of organisms. If the rate of removal or loss of minerals is greater than the rate of addition, the soil will naturally become less fertile. The minerals of the soils are lost due to crops, leaching or soil erosion.

The minerals are often removed from the top layer by rainwater. Cultivation of crops regularly, year after year, makes the soil less productive. In intensive cultivation there are little chances for the restoration of lost nutrients in the soil until they are supplied from outside. The leguminous plants, however, compensate the loss of nitrogenous compounds. Besides this, manure and fertilizers are to be supplemented to restore the fertility of the soil.

The deficiency of mineral nutrients in the soil either can be compensated through organic manures such as green manuring, compost etc. or it can be supplemented by the application of chemical fertilizers from outside sources.

Organic Manures

The organic content of the soil which is a good source of plant nutrients contributes most to the fertility of the soil.

Organic Manures Improve Soil Fertility in the Following Ways

(i)   They modify the physical properties as increase in granulation of the soil and increase in permeability and moisture holding capacity of soil.

(ii) They provide food for soil microbes and thus enhance microbial activities.

(iii) Decomposition products of organic manures help to bring mineral constituents of soil into solution.

(iv) They improve physico-chemical properties of soil, such as cation-exchange and buffering action.

Organic manures are of several kinds some of which are discussed below:

Farmyard Manures

Solid and liquid excreta as dung and urine of all farm animals are termed farmyard manures. They are ready made manures and contain nitrogen, phosphorus and potassium. The farmyard manures of different animals vary greatly in their composition but they are good for all types of soils and all the crops. Farmyard manure when collected in field in exposed condition for several months shows considerable loss of fertilizing value as upon decomposition a considerable amount of ammonia is lost by volatilization.

Therefore, it is important to keep manure protected from weather and manure preparation should be carried out in trenches of about a meter depth. When the trenches are filled with dung etc., the surface is covered with cow dung-earth slurry. In about 3 months the manure becomes ready for use.

Compost

Compost manure can be prepared from a variety of refuse materials, such as straw, sugar cane refuse, rice hulls, forest, litter, weeds, leaves, kitchen wastes. It is prepared in pits usually 6-8in-long, 1½ to 2inwide and one meter deep. In the pits, 30 cm thick layer of plant residues moistened with dung, urine and water is formed and then a second layer of about 30 cm thickness of mixed refuse is spread over it and moistened with slurry. The operation is repeated until the heap rises to a height of about 50 cm above the ground level. The top is then covered with a thin layer of moist earth. After three months of decomposition the material is well mixed and again covered. After a couple of months the manure is ready for use.

There are two types of composts:

(a) Farmyard compost which is obtained from animal excreta and plant residues.

(b) Town compost which is obtained by decomposition of kitchen wastes and garbage of towns and cities. Compost manures are rich in all plant nutrients.

Green Manures

Green manuring is the practice of growing, ploughing and mixing of green crops with soil to improve soil fertility and productivity. Its effects on soils are similar to those of farmyard manures. It is cheap and the best method to increase soil fertility as it can supplement farmyard and other organic manures without involving much cost. Green manures add nitrogen and organic matter to the soil for the improvement of crop yield.

Through green manuring mobilization of minerals, reduction of organic nutrient losses due to erosion, leaching and percolation, and improvement in physical, chemical and biological activities of the soil can be achieved. Green manuring also improves soil aeration and drainage conditions. For green manuring both leguminous and non-leguminous crops are used.

Sawdust

Sawdust can be used as bedding material to conserve animal urine or for making compost. It is a low fertilizing material but it is definitely richer than wheat straw in calcium.

Sewage

In modem system of sanitation, water is used for removal of human excreta and other wastes.

Sewage consists of two components:

(a) The solid part, called sludge and

(b) The liquid part, called effluent or sewage water.

Sewage is quite rich in several plant nutrients and can be used for fertilizing the crop by irrigating the soil directly with sewage water but there is a danger for the spread of several human diseases.

Chemical Fertilizers

Of the elements known to be essential for plant growth, nitrogen (N), phosphorus (P) and potassium (K) are required by plants in pretty large amounts, and are therefore, designated as major ox primary nutrients while calcium, magnesium and sulphur are secondary nutrients. For acid soils, use of Ca and Mg is necessary. Seven elements iron, manganese, boron, molybdenum, copper, zinc and chlorine are required in trace amount and hence called micro-nutrients.

Under continuous cultivation our soils are losing organic matter and mineral nutrients faster than they can be replaced. Regular loss of nutrients from the soil results in compact soil, shallow roots, increased drought, daddy and poorly productive soil. So, for the maintenance of soil fertility quick replacement of the organic matter and mineral nutrients removed from the soils is necessary (Table).

Nitrogen (N), phosphorus (P) and potassium (K), i.e., the primary plant nutrients are commonly applied to soils in the form of commercial fertilizers and hence they are often referred to as fertilizer elements.

Table Fertilizers and their composition

Fertilizer	Composition	Source	Nitrogen %
1. Ammonium sulphate	$(NH_4)_2SO_4$	By-produce from coke and also as synthetic	26.6
2. Ammonium nitrate	NH_4NO_3	Synthetic	33
3. Ammonium chloride	NH_4Cl	Synthetic	25
4. Ammonia liquor	Dilute NH_4OH	Synthetic	20-25
5. Ammonium phosphate	$NH_4H_2PO_4$	Synthetic	11% N and 48% P_2O_5
6. Diammonium phosphate (DAP)	$(NH_4)_2HPO_4$	Synthetic	21% N and 53% P_2O_5
7. Sodium nitrate	$NaNO_3$	Chile saltpetre and synthetic	16
8. Calcium ammonium nitrate	NH_4NO_3 and Dolomite	Synthetic	20.5
9. Calcium cyanamide	$CaCN_2$	Synthetic	21
10. Calcium nitrate	$Ca(NO_3)_2$	Synthetic	15
11. Urea	$CO(NH_2)_2$	Synthetic	45

Chemical fertilizers are classified into the following three group on the basis of materials supplied:

    (i)   Nitrogenous fertilizers.

    (ii) Phosphorus fertilizers, and

    (iii) Potassium fertilizers.

Classification is not as simple as this grouping would imply, several such fertilizer materials as contain two of these elements, as for example, potassium nitrate and ammonium phosphate.

(i) Nitrogenous Fertilizers

Crops usually take nitrogen from the soil in the form of nitrate (NO_3-) and ammonium ions (NH_4+).

Nitrogen fertilizers may be divided for convenience into two groups:

    (a) Organic nitrogen fertilizers, and

    (b) Inorganic nitrogen fertilizers.

Organic Nitrogen Fertilizers

Organic materials such as cotton seed meal, guano and fish tank age are nitrogen carriers but because they supply less than 2% of total nitrogen added in commercial fertilizers, their use is costly and hence they are not used extensively. Nitrogen of organic fertilizers is released slowly by microbial action. They are used as special fertilizers for gardens, lawns and potted plants.

Inorganic Nitrogen Fertilizers

Several inorganic chemicals are used to supply nitrogen to plants. The most important of these are presented in Table.

(ii) Phosphorus fertilizers

Phosphorus has rightly been called 'master key' to agriculture as low crop production is due more often to lack of phosphate than to the deficiency of any other element except nitrogen. In phosphorus fertilizers this element is present in the form of phosphate or superphosphate salts and it is available to the plants when it is combined with organic matter or with calcium and magnesium. Phosphorus is also found in combination with iron and aluminium and is present in certain rock minerals as apatite.

Plants take up phosphorus chiefly as phosphate $(PO_4)-$, HPO_4- and H_2PO_4- ions and the availability of these ions depends chiefly upon the acidity of ground. They become nearly insoluble in strongly acid or strongly alkaline soils. Release of phosphorus from phosphate rocks is slow. The breakdown of phosphate fertilizers produces phosphoric acid (P_2O_5) in soluble form that is absorbed by plants.

Use of phosphate fertilizers on alkaline soils is not suitable. Phosphorus fertilizers are classified into (i) water soluble, (ii) citrate soluble, and (iii) insoluble. When the term P.O, is used it means water soluble plus citrate soluble P2O5. The following are the important phosphate fertilizers being used in all parts of the world including tropical Asia (Table).

Table Phosphate fertilizers

Fertilizer with chemical composition	P_2O_5 available with percentage
(a) Fertilizers containing water soluble phosphorus:	
(i) Superphosphate $Ca(H_2PO_4)_2$ and (ordinary grade) $CaHPO_4$	16-20
(ii) Superphosphate $Ca(H_2PO_4)_2$ and $CaHPO_4$ (concentrated)	40-45
(iii) Ammonium phosphate $NH_4H_2PO_4$	48 (11% N)
(iv) Diammonium phosphate $(NH_4)_2HPO_4$	46-53 (4% N)
(b) Citrate soluble phosphorus fertilizers:	
(i) Dicalcium phosphate	35-40
(ii) Basic slag (Indian) $(CaO)_5P_2O_5SiO_2$	3-5
(c) Insoluble phosphorus fertilizers:	
(i) Rock phosphate Fluor and chlorphatite	20-30
(ii) Bone Meal $Ca_3(PO_4)_2$	18-20

Superphosphate

It is water soluble fertilizer. It does not affect the soil adversely. It contains mono-calcium phosphate, di-calcium phosphate and tri-calcium phosphate, gypsum, silica, iron aluminium sulphate and calcium fluoride and water.

Ammonium Phosphate

It is a fertilizer containing both nitrogen and phosphorus. It is rich in phosphoric acid content but comparatively low in nitrogen content. Ammonium superphosphate (NH4H2PO4). Ca3 (PO4)2 (NH4)2 SO4. It is the cheapest fertilizer and is a mixture of nitrogen and phosphorus fertilizers. It contains nitrogen 3 to 4 per cent and phosphorus pentaoxide (P2O5) 16 to 18%.

Nitro-phosphate

It is highly hygroscopic. It contains nitrogen 13-18 per cent and phosphorus 20 per cent. It is suitable for acid soils.

Bone Meal

It is derived from bone. Bone ash and bone char are the bone products. It is suitable for acidic soils.

Basic Slag

Basic slag, a by-product in the manufacture of steel, is one of the cheapest sources of phosphorus. It is a double compound of silicate and phosphate of lime. It is dark brown in colour and alkaline in reaction.

Rock Phosphate

It occurs in natural deposits. It is light grey in colour. It is a very cheap fertilizer suitable for acid soils.

(iii) Potassium Fertilizers

These fertilizers are soluble in water which means that potassium is readily available to plants. Total potassium of potassium fertilizers is usually expressed in terms of water soluble potassium (K) or potash (K2O). Soils of arid and semiarid areas are generally well supplied with potassium. Acid soils usually need potassium fertilizers more than neutral or alkaline soils because acid soils develop in the areas of high rainfall that leaches out available potassium. All potassium fertilizers are physiologically neutral in reaction.

The following are some common potassium fertilizers:

Potassium Chloride

It is also called muriate of potash. It contains 48—62% K2O. It is also cheap and neutral in reaction.

Potassium Sulphate

It contains about 50% potash (K2O). It is expensive fertilizer.

Kainite

It is natural potassium mineral which contains 14—20% potassium. It is suitable for alkaline soils.

Wood Ashes

It is used in the form of ash as manure. Potassium occurs in the form of potassium carbonate and the percentage of potash is from 2 to 6.5. It is suitable for alkaline soils.

Application of Micronutrients

In order to correct the deficiency of micronutrients, especially if it is very necessary, micronutrients should be added only after ascertaining the amount required. Copper, manganese, iron zinc is supplied generally as their sulphate and boron is applied as borax. Molybdenum is supplied as sodium molybdate. In recent years there has been an increase in use of chelates to supply iron, zinc, manganese and copper.

Soils vary in their ability to supply available nutrients. Some soils may be deficient in nitrogen, some may be deficient in nitrogen and phosphorus, and still some others may be deficient in nitrogen, phosphorus and potassium. To suit the variable requirements of different soils and crops, fertilizer mixtures are prepared. Fertilizer mixtures or mixed fertilizers contain two or more fertilizer materials. If the ingredients and their amounts are known, the formula is referred to as open and if they are not disclosed the formula is termed closed one.

The kinds and amounts of fertilizers to be applied to soil are determined considering the following points:

(a) Kinds of crop to be grown—particularly its economic value, nutrient removal and absorbing ability.

(b) Chemical condition of soil in respect of total nutrients and available nutrients.

(c) Physical state of the soil, especially its moisture content and aeration. For recommending the kind and amount of fertilizers or soil amendments, the analysis of soil is essential.

Application of Soil Conservation Practices

Loss of plant nutrients and water from the soils due to soil erosion can be checked effectively and the fertility of soil can be maintained by application of various biological and engineering methods of soil conservation.

Water Management Practices (Irrigation and Drainages)

Water supply is critical factor in crop production in most areas of the world. Soil moisture greatly affects the availability of mineral nutrients in the soil. It has been proved beyond doubt that fertilizer response is much higher with adequate irrigation.

Drainage and moisture control influence micronutrient availability in soils. Improving the damage of acid soils encourages the formation of less toxic oxidized forms of iron and manganese.

Prevention and Elimination of Inorganic Chemical Contamination of Soil

Loss of soil fertility due to application of toxic chemicals as pesticides can be eliminated if:

(i) Application of toxic chemicals to soil is reduced and

(ii) The soil and crop are so managed as to prevent cycling of toxic chemicals.

Stabilization of Soil pH

The stabilization of pH through application of soil amendments and buffering seems to be an effective guard against the problems of non-availability of certain plant nutrients and radical changes in microbial activities arising due to change in soil pH.

Irrigation Systems

Water is a very important natural resource, which is the basis of all life forms. For sustainable agricultural production, water is one of the most precious important inputs. Plants need water in huge amount throughout their life. Water is also one of the main factors that influence most of the metabolic process such as photosynthesis, respiration, adsorption, opening and closing of stomata and translocation of food material. The growth and yield of crop plants is very much affected by the availability of water.

Water is Used in two Ways

1. Withdrawal or off-channel use.

2. Non-withdrawal or on-channel use.

The amount of water taken out of a streams or pumped out from underground water reservoir or surface water reservoirs to be supplied to the points of major use, such as public water supply systems, irrigation and industries, is referred to as off channel use. In non-withdrawal use, water is used without being removed from its natural source such as for navigation, swimming bodies, wildlife habitats and other recreation purposes.

Need for Irrigation

Irrigation is the artificial watering of soil to sustain plant growth. Irrigation has become necessary because of the limitation of using natural rainfall as the reliable source of water for agriculture. It is also difficult to store the rainwater for immediate irrigation purposes.

The efficient utilization of water resources is essential for better crop production. It includes the suitability of land and water for irrigation, planning of crops and suitable water management practices. Water management includes irrigation and drainage.

The Suitable Irrigation Depends on

- Time of irrigation

- Amount of irrigation, and

- Efficient method of irrigation

Similarly, Suitable Drainage Depends on

- How much to drain

- How best to drain

- How rapidly to drain

Stages of Crop when Irrigation is Required

The growth span of a crop plant passes through various phases and stages of growth. The rate of irrigation varies during the different stages of plant growth, i.e., from seedling to maturity stage. The growth period of irrigated crops can generally be divided into three phases, namely vegetative, flowering and maturity stage. At vegetative stage, light and intermittent irrigation is required whereas at flowering stage moderate and frequent irrigation is needed and during crop maturation stage again light irrigation is required.

Systems of Irrigation

There are different water sources from which irrigation is carried out, for example, canals, wells, open wells, tube wells, tanks etc.

Depending upon the crops, soil types, water resources, climate conditions and costs involved, several systems of irrigation are used which are as follows:

1. Surface irrigation systems

2. Subsoil irrigation systems

3. Drip irrigation systems

Surface Irrigation System

In this system, water is directly used on the surface of the soil and water follows the slope of the land, the surface irrigation system may be of following types:

*   Flood Irrigation:

    Flood irrigation is used for close-grown crops such as rice and where farm fields are leveled and water is abundant. A sheet of water is allowed to advance from different sources and remains on a field for a given period, depending on the crop, the porosity of the soil and its drainage.

Fig. Different surface irrigation systems: (a) Flood irrigation (b) Furrow irrigation (c) Basin irrigation

*   Basin Flooding:

    It is used in widely spaced trees, such as in orchards, with basins built around trees and filled with water (Figure).

*   Check Basin:

    This system of irrigation is highly popular as it is more suitable to all types of soils and to a variety of crops.

*   Furrow Irrigation:

    In furrow irrigation, water moves in the fields in furrows, between two ridges. It is employed in the fields with row crops such as cotton and vegetables. Parallel furrows, called corrugations, are used to spread water over fields that are too irregular to flood.

Sprinkler Irrigation System

In this system, irrigation is done through pressure, to the surface of any crop or soil, in the form of a thin spray. Sprinkler irrigation system uses less water and provides better control. Each sprinkler, spaced along a pipe, sprays water in a continuous circle until the moisture reaches the root level of the crop. Centre-pivot irrigation uses long lines of sprinklers that revolve in a circular field. It is used especially for feed crops such as alfalfa, tall crops and orchards.

1. Sprinkler irrigation system conveys water from the source to the field, through pipes under pressure, and distributes over the field in the form of spray of 'rain like' droplets.

2. 10-20% of area is not irrigated at the comers of square of rectangular plot.

3. This system requires high energy and involves huge cost of the equipment.

This system can be followed in conditions when:

1. The soil is too shallow.

2. The land is too steep.

3. Less and frequent irrigations are needed.

4. The soils are very sandy.

This system is disadvantageous in the following conditions when:

1. Strong winds cause improper distribution of water.

2. Evaporation losses are high from sprinkler irrigation, especially under high temperature and low relative humidity conditions.

3. The initial cost is high.

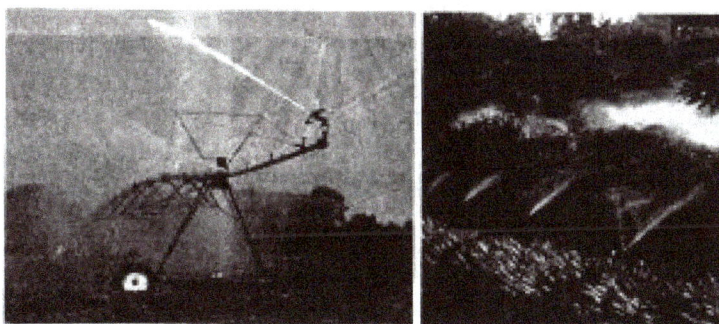

Fig. Sprinkler irrigation: a. center-pivot irrigation with long lines of sprinkler,
b. sprinkler irrigation system in which water is conveyed to field through pipes under pressure.

Sub-Soil Irrigation System

In this system, irrigation is done into a series of ditches in the field deeply to the impervious layer. Then it moves laterally and vertically through capillaries saturating the root /one. In this system, continuous supply of water in the root zone is assured from the artificial water table created by the ponding of irrigation water on the impervious layer. This system is very efficient because the water losses through evaporation from surface can be reduced. This system is more common in Gujarat and Jammu and Kashmir, for cash crops growing on sandy loam soils.

Drip Irrigation System

It is also called trickle irrigation. This irrigation system involves the slow application of water, drop by drop, to the root zone of a crop. It is done through mechanical devices called emitters, located at selected points along water delivery lines. Drip irrigation is extensively used in areas of acute water scarcity and especially for crops such as coconut, grape, banana, citrus, sugarcane, cotton, maize, tomato and plantation crops.

Fig. 26.3 Drip irrigation

Fig. Drip irrigation

Advantages of Drip Irrigation

1. Water loss through transpiration is low.

2. It is possible to obtain better yield and quality of crops by controlling soil moisture-air nutrient levels.

3. We can save the fertilizers by monitoring the supply of nutrients as per the need of the crop.

4. Improvements in biological fertility can be achieved by avoiding pollution.

Disadvantages of Conventional Irrigation Methods

i.   Major part of the water goes waste; only small quantity of water is utilized by the plants.

ii.   Water is not uniformly distributed due to uneven and poor leveling off the field.

iii. Crops are usually subjected to cyclic changes of flooding and water stress situations by providing heavy irrigation at one time and leaving the fields to dry up for about 10 to 15 days.

iv.   The low lying fields always get excess water that causes prolonged water logging, due to lack of leveling of fields.

v. In the fields, about 10-15% of land is utilized for making channels and distributaries etc. which decreases the area of cultivation.

vi. Excessive irrigation and poor water management leads to water logging and accumulation of salty in the upper layer of the soil.

vii. In conventional irrigation, unmanageable and undesirable weeds grow.

Problems of Excess Irrigation

i. Excess irrigation causes several changes in the soil and plants, resulting in reduced growth and sometimes even death of plants.

ii. Germinating seeds are sensitive to water logging, since they are totally dependent on the surrounding soil space for oxygen supply.

iii. Yield of cereals is reduced if excess irrigation is given.

iv. Excess water causes injury to the plant due to low oxygen supply to the root system and accumulation of toxic substances in the soil.

v. Leaching of nitrates and de-nitrification occur which result in nitrogen deficiency.

vi. Permeability of roots decrease due to shortage of oxygen. It results in decreased uptake of water and nutrient.

Chapter 2

Classification of Soil

In order to understand the suitability of a particular soil or to determine the relationship between varied soils, the soil is classified into different categories. The topics elaborated in this chapter include clay soil, sandy soil, red soil, silt soil, topsoil, alkali soil, etc. which will help in developing a better perspective about the classification of soil.

Clay Soil

Clay is another variation of soil used in the world of viticulture for growing high-quality grapes. There are quite a few types of clay soils, each with its own set of characteristics. Calcareous, Marl and Albarese are very well-known types of limestone-based Clay soils. Clay soils tend to retain a lot of water and are recognized for their superior nutrient and mineral properties. Some of the world's boldest wines are made from grapes grown in Clay soil.

Clay soil is made up of tiny particles that tend to store water for prolonged periods of time. This increases the tendency of the soil to remain cooler, the grape vines greatly benefit from this aspect in harsh weather conditions and when they are not watered.

The famous red wines produced from Clay soils include Cabernet Sauvignon from Rioja, Spain, the highly acclaimed Pinot Noir from Burgundy, the muscular Sangiovese from Chianti and last but not least the widely recognized Shiraz from Barossa, Australia.

Though different soils have a wide range of colors, textures and other distinguishing features, there are only three types of soil particles that geologists consider distinct. The quality of soil depends on the amount of sand, loam and clay that it contains, because soils with differing amounts of these particles often have very different characteristics. Soil with a large amount of clay is sometimes hard to work with, due to some of clay's characteristics.

Particle Size

Clay has the smallest particle size of any soil type, with individual particles being so small that they can only be viewed by an electron microscope. This allows a large quantity of clay particles to exist in a relatively small space, without the gaps that would normally be present between larger soil particles. This feature plays a large part in clay's smooth texture, because the individual particles are too small to create a rough surface in the clay.

Structure

Because of the small particle size of clay soils, the structure of clay-heavy soil tends to be very dense. The particles typically bond together, creating a mass of clay that can be hard for plant roots to penetrate. This density is responsible for clay-heavy soil being thicker and heavier than other soil types, and clay soil takes longer to warm up after periods of cold weather. This density also makes clay soils more resistant to erosion than sand or loam-based soils.

Organic Content

Clay contains very little organic material; you often need to add amendments if you wish to grow plants in clay-heavy soil. Without added organic material, clay-heavy soil typically lacks the nutrients and micronutrients essential for plant growth and photosynthesis. Mineral-heavy clay soils may be alkaline in nature, resulting in the need for additional amendments to balance the soil's pH before planting anything that prefers a neutral pH. It's important to test clay-heavy soil before planting to determine both the soil's pH and whether it lacks important nutrients such as nitrogen, phosphorus and potassium.

Permeablity and Water-Holding Capacity

One of the problems with clay soil is its slow permeability resulting in a very large water-holding capacity. Because the soil particles are small and close together, it takes water much longer to move through clay soil than it does with other soil types. Clay particles then absorb this water, expanding as they do so and further slowing the flow of water through the soil. This not only prevents water from penetrating deep into the soil but can also damage plant roots as the soil particles expand.

Identifying Clay

There are several tests you can use to identify clay soils. If rubbed between the fingers, a sample of clay soil often feels slick and may stick to the fingers or leave streaks on the skin. Rubbed clay soil often takes on a shiny appearance as well, as opposed to the rough texture you would see with other soils. Clay soils do not crumble well, and a sample of clay can typically be stretched slightly without breaking. When wet, clay soils become slick and sticky; the soil may also allow water to pool briefly before absorption due to the slow permeation. Visually, clay soils seem solid with no clear particles, and may have a distinct red or brown color when compared to the surrounding soil.

Clay soil has several uses for around the home, the art studio and even the science lab. The soil content is well suited for several species of trees and shrubs that thrive on the nutrient-rich top layers of the clay, while moist, compacted portions of the soil have played a key role in ceramics and pottery for thousands of years. Additionally science has begun to look at the elements within clay soil as doctors and scientists look at clay as more than a holistic remedy for illnesses but perhaps a real effective disease fighter.

Tree Planting

Clay soil is well tolerated by tree populations with some species actually preferring the soil variety over sand or loam grades. According to bachmans.com, willow, aspen, cottonwood, Ohio buckeye, hawthorn and three species of ash grow best in well-ventilated, moist clay soil.

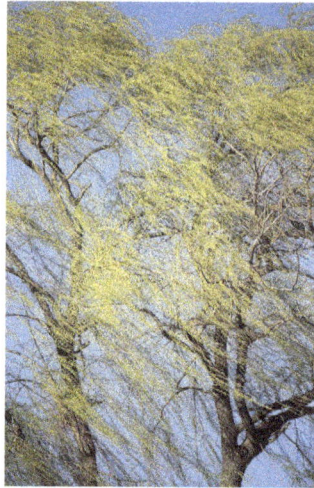

Medicinal

Minerals found in clay soil have been used in folk medicine to relieve an upset stomach and prevent wound infection. According to a 2008 study published in the Journal of Antimicrobial Chemotherapy, clay minerals rich in iron were effective at fighting antibiotic-resistant strains of pathogenic bacteria in a lab setting. The study concluded that specific clay soil minerals have antibacterial properties that could produce an inexpensive means of fighting infection in humans.

Pottery

One of the more popular uses of clay soil is in ceramics and pottery. The soil is easily combined with water and worked into a thick mass that is well suited for sculpting. Clay must be kept in a dark, moist environment in order to keep from drying out and to retain its ability to be molded into different shapes. The molded clay is then fired in an oven to remove the moisture from the pot. This process takes several hours to complete, as the clay must be heated slowly in order to keep moisture from evaporating too quickly and causing cracks in the pottery.

Sandy Soil

Sandy soil is easy to spot by its feel. It has a gritty texture and when a handful of sandy soil is squeezed in your hand, it will easily fall apart when you open your hand again. Sandy soil is filled with, well, sand. Sand is primarily small pieces of eroded rocks.

Sand tends to have large particles and the particles are solid and have no pockets where water and nutrients can hold to it. Because of this, water and nutrients tend to run out, and because sandy soil lacks both water and nutrients, many plants have a difficult time surviving in this kind of soil.

Improvement of Sandy Soil

The best sandy soil amendments are ones that increase the ability of the sandy soil to retain water and increase the nutrients in the soil as well. Amending sandy soil with well-rotted manure or compost (including grass clippings, humus and leaf mold) will help to improve the soil the fastest. You can also add vermiculite or peat as sandy soil amendments, but these amendments will only add to the soil's ability to hold onto water and will not add much nutrient value to the sandy soil.

When amending sandy soil, you need to watch the salt levels of the soil. While compost and manure are the best way to amend sandy soil, they contain high levels of salt that can stay in the soil and damage growing plants if the salt level builds up too high. If your sandy soil is already high in salt, such as in a seaside garden, be sure to use plant only based compost or sphagnum peat, as these amendments have the lowest salt levels.

Uses of Sandy Soil

The uses of sandy soil in different sectors are numerous. Followings are the uses of sandy soil:

- Agricultural need.
- Easy drainage
- Construction
- Foundation
- Beauty
- Frictional properties
- Low settlements
- Changing pH
- Others

Agricultural Need

Sandy soil is usually dry, nutrient and fast draining. It is used for plowing, planting and cultivating. The useful vegetables like potatoes, grams, tomatoes etc. require a minimal percentage of soil for a specific period. The percentage varies from vegetable to vegetable. Sandy soil also provides a good ground for farmers to collect falling nuts.

Easy Drainage

Sandy soil has great drainage properties. It drains easily and quickly. It is used to improve soil drainage. The interesting feature is in a land full of sandy soil, work can be done right after a rain even if it's heavy without any difficulty. It filters water in big deposits through which water is shred and recollected through the channels at the bottom. Sandy soil also retains flower water like soaking the water.

Construction

Sandy soil doesn't get sticky. It is cohesion less. It has light and loose structure. That's why it can be easily used for construction purpose. Sandy soil can be a great aggregate for concrete. Also, it can be used as a construction material of mortar aside cement. Sandy soil is also used for the erection of exterior rendering materials. It is used because of its chemical resistance. Sandy soil also can be used the best as filling sand.

Foundation

Sandy soil provides a base to foundations on swampy ground. Densification below foundation is not required as the soil is naturally in a dense state. Sandy soil provides a reliable foundation for which the structure that relies on this soil has a greater degree of reliance than these built-in other soils. Sandy soil is maneuvered for the improvement of ground with the use of soil replacement method. It is used to replace soft clays of foundation to improve bearing capacity of the soil.

Beauty

Sandy soil starts from the almost sandy surface like beaches. It increases the beauty of the beach. Also, it's used in gardening and kids playgrounds for safety by providing a soothing context.

Frictional Properties

Sandy soil has very good frictional properties. The frictional properties are used in the construction of reinforced soil structure with geosynthetics reinforcement.

Low Settlements

Sandy soils have low settlements as it does not undergo consolidation with time. Moreover, it has immediate settlements.

Changing pH

The pH level of sandy soil can easily change the pH level of soil like clay. The pH level of sandy soil is between 7.00 and 8.00.

Others

Sandy oil is used to reduce the velocity of water. It also helps to percolate the soil and raise the water table.

Red Soil

The color of red soil ranges from red to brown, chocolate, yellow, gray, or sometimes even black. Red soil contains a high percentage of iron content, which is responsible for its color. This soil is deficient in nitrogen, humus, phosphoric acid, magnesium, and lime but fairly rich in potash, with its pH ranging from neutral to acidic. It is formed by the weathering of ancient crystalline and metamorphic rocks, particularly acid granites and gneisses, quartzitic rocks, and felspathic rocks. Chemically, red soil is siliceous and aluminous, with free quartz as sand, but is rich in potassium, ranging from sand to clay with the majority being loamy. The lowermost area of red soil is dark in color and very fertile, while the upper layer is sandy and porous. Thus, proper use of fertilizers and irrigation yields high production of cotton, wheat, rice, pulses, millets, tobacco, oil seeds, potatoes, and fruits.

Red soils are formed as the result of the drainage of the rocks and the lime stone, red soils are less clayey and sandier in nature. The red soil contains the very rich content of iron plus small humus content. The content the beneficial material that is essential nutrients like nitrogen, phosphorus, and lime is very less in red soils or it has lacks of that.

This soil contains some mount of the acidic nature and the moisture of the soil contains. Because of lime deficiency and soluble salt content, it is due to the presence of iron oxide deposits, that red soils get the unique red tint and are comparatively infertile.

The presence of a good concentration of iron oxide and the reach in iron is responsible for giving this soil its reddish shade or yellowish. It is in the state of some concentrated and heavy, it consists of a considerable concentration of iron oxides similar to Laterite soil in India.

If it is get compared then it's got to know that, Red Soils are less clayey and sandier. Which receive significantly low rainfall and therefore are less leached in comparison to the Laterite soil in India moreover, Red soils are formed in those areas. Further, red or yellow soils build up generally on metamorphic and the drainage of the rocks.

These soils are acidic in nature it means it have acidic property and this is one similarity between laterite soils and red soils. Red soils are not able to preserve moisture and the wet of the soil and hence, crops are cultivated mostly during the rainy season. Constant application of manures is very vital, in order to get a higher yield in this type of soil.

Red soils mostly developed and find in the Great Indian Peninsular Plateau. Amazingly in the valleys and lowlands, Red Soils are deep and very fertile because it gets water regularly. On the other hand, slopes of the hills and mountain, they are mostly poor and thin.

Silt Soil

Silt is somewhere between the size of sand and clay, and is an important component in the sedimentary dynamics of rivers. Silt comes in several forms. It might be found in the soil underwater or as sediment suspended in river water. Silt is geologically classified by its grain size and texture going through a sieve. Letters are assigned to the grain of soil, whether it is gravel, sand, silt, clay, or organic. Then, it is further delineated as to whether the sample is poorly graded, well-graded, has a high plasticity, or low plasticity. The sample composition is determined by passing it through differently sized sieves, and the result is classified with the combination of letters assigned to it based on its physiochemical characteristics.

Chemical Composition

Silt is an aggregation that comes mainly from feldspar and quartz, although some other minerals could also be part of its composition. The erosion of these source minerals by ice and water starts the transformation that eventually turns these broken minerals into silt that are no more than .002 inches across. Silt, sand, clay, and gravel all mix to form soil. Silt is also determined by the naked eye and touch by its slippery, non-sticky feel when wet, as opposed to clay, sand, or gravel. It has a flour-like consistency when dry. Silt is found more in semi-dry environments than anywhere else.

Role in Agriculture

In its dry state, silt is a fine powdery substance that is much like dust. It is easily carried in the air by wind, and may be transported many miles away. This enables silt to convert unproductive land into fertile land as it is deposited. This is similar to desert dust that settles in ocean waters and rivers that becomes part of the ecosystem. This action enriches the waters and agricultural lands, much like the ancient annual flooding of the Nile River and the Mississippi River Delta siltation in the United States that in turn supported the rich harvests of crops. "Silt stones" can also be produced by the compression of silt deposits. Silt stones have building and garden uses due to their light weight. It is also used to make mortar and natural cement, as well as in soil conditioners.

Harmful Impacts of Silt

Siltation occurs as a result of human activities that lead to fine soil leaching into nearby rivers. This results in an unnaturally large accumulation of silt that stays in that particular area of that river. Rainstorms may also transport these soils into other water sources. Sensitive marine life and freshwater fish may be affected by suspended silt in their native waters. Benthic organisms such as coral, oysters, shrimps, and mussels are especially affected by silt, as they are filter feeders that may literally become "choked up" by silt-laden waters. Waterways and irrigation canals could also become affected in their functions by silt accumulations. Other harmful impacts of siltation are human health concerns, the loss of wetlands, coastline alterations, and changes in fish migratory patterns.

How Human Practices Affect Natural Sedimentation

Human activities all contribute to enrich natural sedimentation and cause sediment deficiency in rivers, lakes, and oceans. Some sources of increased sedimentation are given rise to by construction activities that require the clearing of land, river dredging, offshore dumping, and climate change. These activities all contribute to the pollution and degradation of aquatic resources. On the other hand, sedimentation control can be achieved by improved land use practices such as modified cropping, terracing, low tillage farming, creating buffer zones, and wetland conservation. Sedimentation is a crucial component of life in an aquatic ecosystem, but sedimentation can also damage the aquatic ecosystem.

Peat Soil

Peat soils are formed from partially decomposed plant material under anaerobic water saturated conditions. They are found in peatlands (also called bogs or mires). Peatlands cover about 3% of the earth's land mass; they are found in the temperate (Northern Europe and America) and tropical regions (South East Asia, South America, South Africa and the Caribbean).

Peat soils are classified as histosols. These are soils high in organic matter content. Peat formation is influenced by moisture and temperature. In highly saturated anaerobic soils, decomposition of plant material by microorganisms is slowed down, resulting in high carbon accumulation. In colder climates decomposition of plant material by microorganisms is slowed down leading to quicker peat formation. The carbon content of peat soils makes peatland a major storage of carbon on the earth surface. This is why its importance in fighting climate change can never be overemphasized.

Some Economic Benefits of Peatlands

Peatlands bring enormous economic benefits to regions where they are found.

- Peat is extracted for use as horticultural compost. It is highly sought after in commercial horticulture because of its high water retaining ability and flow of air.

- Peat is used for fuel to generate electricity. It is also sold as briquettes for heating homes in cold climatic regions.

- Peatlands are drained and used for agricultural purposes (pasture and crop production) and forestry.

Peat use for forestry and agriculture are beneficial but it alters the natural peatland hydrology. This causes oxidation of stored carbon therefore declining its organic matter content. During peat extraction, peat is drained and dried before storage or transportation for sale. These processes reduce the water content and encourage microbial decomposition of organic matter. The result of this is the release of greenhouse gasses such as CO_2 and N_2O.

Consequences of Peat Disturbance

Apart from greenhouse gas emission, peatland disturbance brings a number of other changes:

- Drainage of peatland causes decline in biodiversity because its natural hydrological habitat is disturbed. Peatlands provide habitation for diverse species of meadow birds, animals, vegetation and insects.

- Peat oxidation can lead to release of dissolved organic matter and peat particles into surface waters.

- Peat oxidation can lead to the loss of a historical heritage. Peat soils have the ability to store human remains or ancient artefacts for thousands of years; since they have very minimal microbial decomposition. A good example of this is the 4000 year old body of a man found in peat from Cashel-Central Ireland.

- Peat soils drained for agricultural purposes are more vulnerable to wind and water erosion when the topsoil is severely dry.

- Drainage of peatland can lead to peat fires which destroy forestland and habitation and further increase the emission of CO_2 to the atmosphere.

Although the above listed are negative consequences of peat disturbance, it is good to acknowledge that when peat soils are drained for use in agriculture, decomposition of organic matter is accelerated leading to the mineralization of nitrogen (a vital nutrient for plant growth). This in fact is a good thing. You can learn more about this from the nitrogen cycle.

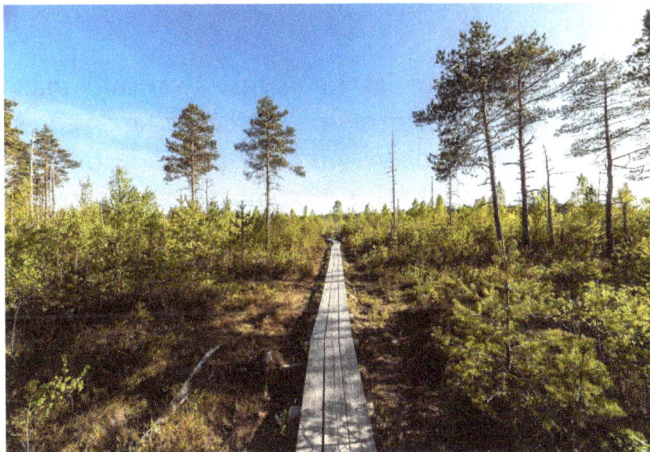

Peatland Conservation and Restoration

1. Conserve wet peatlands: This approach is preventive and avoids the expensive cost of restoring peatlands to their natural hydrological state. This is simply putting a stop to the drainage of peatlands. There is no need for soil restoration projects if efforts are made to keep the soil in its

natural state. People in surrounding communities must be educated on the benefits of conserving the peatland natural ecosystem.

2. Use of paludiculture: This method also maintains the wetness of peatlands. Paludiculture involves the cultivation of biomass crops on peatlands without disturbing the peat natural hydrology or ecosystem. There is no drainage of the peatland involved. It has multiple benefits; it reduces peat oxidation, greenhouse gas emission and at the same time supplies biomass used for combustion. It also serves as a source of food for neighboring communities as some edible crops grow on the wetland. Examples of edible crops in paludiculture are wild rice -Zizania aquatic (also called floating rice), wild edible berries (blue berries, black currant, black raspberries) and sweet grass Hierochloe odorata. Paludiculture is practiced in Europe (Russia and Belarus) North America and Asia (Indonesia and Malaysia).

3. Re-wetting and restoration: In re-wetting effort is made to restore the soil back to its natural hydrological, anaerobic state by raising the water table level to the land surface. Re-wetting reduces CO_2 and N_2O emissions but increases the emission of CH_4 (which is released naturally in undisturbed peatlands). Under anaerobic conditions decomposition of plant material by microorganism is slow but still in action. Decomposition of organic material under this condition is carried out by methanogenic Archaea (a methane producing microorganism).

Chalky Soil

Chalky soil is comprised mostly of calcium carbonate from sediment that has built up over time. It is usually shallow, stony and dries out quickly. This soil is alkaline with pH levels between 7.1 and 10. In areas with large deposits of chalk, well water will be hard water. An easy way to check your soil for chalk is to put a small amount of the soil in question in vinegar, if it froths it is high in calcium carbonate and chalky.

Chalky soils can cause nutrient deficiencies in plants. Iron and manganese specifically get locked up in chalky soil. Symptoms of nutrient deficiencies are yellowing leaves and irregular or stunted growth. Chalky soils can be very dry for plants in the summer. Unless you plan to amend the soil, you may have to stick with drought tolerant, alkaline loving plants. Younger, smaller plants also have an easier time establishing in chalky soil than larger, mature plants.

Chalky soils are highly variable and range from gravelly to clay-like. The clay-like element may in fact be mainly finely divided calcium carbonate making it very poor for plant growth. However where true clay is present in the soil, nutrient levels may be higher and the water holding capacity greater.

Chalky soils can be identified by:

- Chalky or lime-rich soils may be light or heavy but are largely made up of calcium carbonate and are very alkaline (they have a pH of 7.1-8.0).

- If soil froths when placed in a jar of vinegar, then it contains free calcium carbonate (chalk) or limestone and is lime rich.

- Very chalky soils may contain lumps of visible chalky, white stones and often large sharp flints which can easy split. Limestone soils contain lumps of limestone.

Gardening with Chalky Soil

Naturally lime-rich soils contain chalk and limestone in excess, and are often associated with down land, herb-rich pastures and chalk and limestone woodlands.

Light chalky soils are:

- Often full of stones.

- Can be severely dry in summer.

- Often poor in nutrients as both manganese and iron can be "locked up" in the soil so that they become unavailable to plants.

- Shallow and light (but easier to cultivate because of their weight and looser construction than "heavy" clay).

But on the plus side:

- Flooding is rare on light alkaline soils because of their elevation and porosity.

- They warm up quickly in spring.

- With good manuring and fertilizer that can be moderately fertile and ideal for growing a wide range of plants.

- Brassicas are less likely to suffer from club root in chalky soils.

Techniques and tips if you garden on light chalky soil:

- When wet, chalky soils may become sticky and unworkable, but they can be cultivated soon afterwards without doing serious damage (unlike clay soils).

- Smaller plants get established more quickly than more mature specimens.

- Mediterranean and prairie plants should grow well on chalky soils.

- Avoid acid-loving plants such as camellias and rhododendrons unless you are prepared to grow them in containers.

Improvement Chalky Soil

Five Steps to Improving Chalky Soils

1. Dig in plenty of organic matter to help improve moisture retention and humus levels in the soil (this can break down and disappear very quickly).

2. Chalk and limestone subsoils are usually deeply fissured and tree and shrub roots find moisture readily, but in some cases it may be necessary to break up the sub-soil to get

sufficient depth to plant woody plants.

3. Apply fertilizers.

4. Mulch plants with organic matter to conserve moisture.

5. Green manure such as crimson clover (Trifolium incarnatum), vetch and bitter blue lupin (Lupinus angustifolius) can be grown to help fix nitrogen in the vegetable garden and before planting elsewhere.

Loamy Soil

Loam soil is considered the ideal planting medium for growing a wide variety of plants. Heavy clay soil has poor drainage and can often be hard to work, while sandy soils fail to properly retain water and nutrients that plants require for growth. Loam soil is a fairly equal combination of three soil types: sand, silt and clay. Loam drains well, yet holds moisture and nutrients needed for growing healthy plants. The medium-textured soil also allows air to circulate around plant roots while providing protection from diseases often caused by other types of poorly drained compacted soils.

Raising Crops

Loam soil provides vegetable plants with soil conditions needed to produce abundant crops during the growing season. Choose a vegetable planting location that receives six to eight hours of daily sunlight. Amend clay or sandy soil prior to vegetable planting with soil conditioners like compost, manure, sawdust, peat moss or coarse sand. Add 3 to 4 inches of organic materials and 1 to 2 inches of coarse sand to the soil surface, then mix in with a garden tiller or spade 8 to 10 inches into soil. To prevent nutrient deficiency caused by decomposing organic matter, add nitrogen when necessary.

Planting Trees

Young trees prefer being grown in loam because the texture of the soil allows roots to spread quickly, retain moisture and obtain nutrients needed to grow. Select a planting location with the proper daily light requirements for your tree. Dig a hole two to three times larger than the tree root ball, and then mix conditioners into extracted soil to amend clay or sandy soil to a loamy consistency. Back fill with the new soil mixture, keeping trees at their original planting depth. If amended sandy soil does not hold new transplanted trees well, add a layer of heavier topsoil and mulch to the surface.

Growing Flowers

Many types of flowers grow quickly and thrive in loamy soil conditions. Pick garden sites that provide full sun to partially shaded conditions as required by your plants. Adjust heavy clay or sandy soil with conditioners as necessary using a garden tiller or hoe. Plant annual or perennial flower transplants into amended soil and add a 2-inch layer of organic mulch to the surface around the plants. Fertilize with an all-purpose, water-soluble fertilizer every seven to 10 days throughout the warm spring and summer months to produce the greatest amount of colorful flowers.

Topsoil

Topsoil is defined as the 2-8 inches of soil on the surface of the Earth. The reason why it is called 'topsoil' is because it is on the uppermost layer of the dirt that you work with when gardening or seeding. Subsoil, or the layer directly below topsoil, consists of hard-packed dirt and clay. It is largely unsuitable for anything organic except for large tree roots.

Topsoil is one of those things that every yard needs, but for some, they never really know what it is made of or how best to use it. Topsoil is the first line of defense for growing grass and plants, as well as fending off herbivore pests. Understanding the composition of topsoil, as well as what some of the best uses for it, is important to maintaining the health of the lawn and/or garden look its best.

The Components of Topsoil

Topsoil is an important part of any kind of lawn or garden because of its contents. Topsoil has high amounts of organic matter as well as microorganisms. As plants, vegetables, and other organic matter dies, it slowly decays, and the nutrients are returned to the topsoil from where it came. Microorganisms, as well as organic matter, can help greatly with growing and keeping out bacteria that may cause issues with plants.

Other elements of topsoil include sand, manure of various types and very small pieces of rock and stone. Its composition varies based on location, for topsoil is not the same everywhere.

Topsoil can be found at the local gardening center. Often, topsoil can be purchased as a mix of sand, silt, and clay. The difference between the topsoil that you will find at the home is that the topsoil at the gardening centers gives you the exact percentages of what you can find in each bag of topsoil.

When just using the topsoil around the home, often the components are in the incorrect proportions. So, dig up some of the topsoil and test it to see what percentage of the different components already exists in the soil. It is as simple as going to the local hardware store, finding a soil test kit, and following the instructions. Often you will be instructed to dig a small hole in the parts of the yard that you are planning on testing, and then the kit will determine the pH balance of the soil. Knowing what kind of soil you already have will help you figure out what kind of topsoil to purchase at the local garden center for more conducive planting.

Uses Topsoil

Topsoil is best used in the yard to allow nature to do the heavy-lifting for you. In other words, a high-quality topsoil composition is what you need to encourage healthy growth for plants, trees, vegetables, fruits, and shrubbery. Topsoil can also prevent weeds and pests from infecting the plants.

Using store-bought topsoil can protect bulbs and root systems, as well as the buds still growing underneath the soil. Topsoil is much less expensive than garden soil, which is usually richer in nutrients. So, when you use the topsoil, you can use it more freely than other soils.

Topsoil often comes in larger bags, so make sure that you are getting the right kind before you purchase. Otherwise, you may have a large bag of something that will just sit in the back of the garage or shed.

Topsoil can be a bad idea when you want something to blanket the entire yard or gardening area. It is made to be used in large chunks but is supposed to work with the yard rather than be a replacement for it. The naturally occurring topsoil under the feet has come from millions of years of the Earth trying to figure out what works best, so trying to use all store- bought topsoil that isn't a native composition to the yard for is probably not a good choice. Topsoil is there to enhance what is already there or to help plants struggling to grow due to an incorrect balance. So, when making use of topsoil makes sure that it used correctly and sparingly.

Alkali Soil

Alkaline soil is any soil that falls above 7 on the pH scale - the unit of measure for acidic versus alkaline soil. The pH scale is a numeric system between 0 and 14 used to test the acidity or alkalinity of soil with a measurement of 7 being completely neutral.

Soils with a high pH tend to contain higher levels of sodium, calcium and magnesium. The availability of nutrients is often limited and plants can become stunted in alkaline soils because they're not as soluble as acidic soils.

Alkaline soil may also be known as sweet soil.

Knowing the pH of the soil is important. Periodic testing of both the topsoil and subsoil with a pH testing kit purchased from a hardware or garden store will give an accurate measurement of the alkalinity.

Alkaline soil can occur naturally in places with little rainfall. For example, the sandy soil of deserts is alkaline. Dense forests, peat bogs and soil high in certain minerals are high alkaline soils. Hard water that includes lime can also raise the pH of soil to alkaline levels.

Most often lime or peat moss is added to the soil to increase the alkalinity of soil.

Some plants, such as lilies, geraniums and maiden hair fern thrive in alkaline soil.

Serpentine Soil

Serpentine soils are weathered products of a range of ultramafic rocks composed of ferromagnesian silicates. Serpentine more accurately refers to a group of minerals, including antigorite, chrysotile, and lizardite, in hydrothermally altered ultramafic rocks. Common ultramafic rock types include peridotites (dunite, wehrlite, harzburgite, lherzolite) and the secondary alteration products formed by their hydration within the Earth's crust, including serpentinite, the primary source of serpentine soil. Serpentine soils are generally deficient in plant essential nutrients such as nitrogen, phosphorus, potassium, and sulfur; have a calcium-to-magnesium (Ca:Mg) molar ratio of less than 1; and have elevated levels of heavy metals such as nickel, cobalt, and chromium. Although physical features of serpentine soils can vary considerably from site to site and within a site, serpentine soils are often found in open, steep landscapes with substrates that are generally shallow and rocky, often with a reduced capacity for moisture retention. Due to the intense selective pressure generated by such stressful edaphic conditions, serpentine soils promote speciation and the evolution of serpentine endemism, contributing to unique biotas worldwide, including floras with high rates of endemism and species with disjunct distributions. The biota of serpentine soils has contributed greatly to the development of ecological and evolutionary theory, as well as to the study of the genetics of adaptation and speciation. Plants growing on serpentine soils also provide genetic material for phytoremediation and phytomining operations. Habitats with serpentine soils are undergoing drastic changes due to ever-expanding development, deforestation, mining for heavy metals and asbestos, exotic-species invasions, climate change, and atmospheric deposition of previously limiting nutrients such as nitrogen. Such changes can have drastic impacts on serpentine floras and affect bacteria, fungi, and fauna associated with serpentine plants and soils. Habitats with serpentine soils provide ample opportunities for conservation- and restoration-oriented research directed at finding ways to better manage these biodiversity hotspots.

References

- Amending-sandy-soil, garden-how-to, soil-fertilizers: gardeningknowhow.com, Retrieved 22 June 2018

- Properties-clay-soil-71840: homeguides.sfgate.com, Retrieved 31 March 2018

- Red-soil, agricultural-and-biological-sciences, sciencedirect.com, Retrieved 12 July 2018

- What-is-silt-and-how-does-it-impact-the-environment: worldatlas.com, Retrieved 22 April 2018

- Uses-loam-soil-123885: gardenguides.com, Retrieved 28 June 2018

- Alkaline-soil-8: maximumyield.com, Retrieved 11 July 2018

- Serpentine-Soil-272820510: researchgate.net, Retrieved 12 March 2018

Chapter 3

Pedology

Pedology is the science that is concerned with the classification, description and formation of soils. It is a branch of soil science. This chapter discusses in detail the significant aspects and principles of pedology. It includes vital topics such as pedogenesis, soil morphology, active layer and soil map.

Pedology is a scientific discipline concerned with all aspects of soils, including their physical and chemical properties, the role of organisms in soil production and in relation to soil character, the description and mapping of soil units, and the origin and formation of soils. Accordingly, pedology embraces several sub disciplines, namely, soil chemistry, soil physics, and soil microbiology. Each employs a sophisticated array of methods and laboratory equipment not unlike that used in studies of the physics, chemistry, or microbiology of non-soils systems. Sampling, description, and mapping of soils is considerably simpler, however. A soil auger is used to obtain core samples in places where no subsurface exposure can be found, and the soil units are defined, delineated, and mapped in a manner similar to procedures in stratigraphy. Such soils studies, in fact, overlap the concerns of the stratigrapher and the geologist, both of whom may treat the soils layers as strata of the Quaternary Period (from 2.6 million years ago to the present).

Soil is not only a support for vegetation, but it is also the zone beneath our feet (the pedosphere) of numerous interactions between climate (water, air, temperature), soil life (micro-organisms, plants, animals) and its residues, the mineral material of the original and added rock, and its position in the landscape. During its formation and genesis, the soil profile slowly deepens and develops characteristic layers, called 'horizons', while a steady state balance is approached.

Soil users (such as agronomists) showed initially little concern in the dynamics of soil. They saw it as medium whose chemical, physical and biological properties were useful for the services of agronomic productivity. On the other hand, pedologists and geologists did not initially focus on the agronomic applications of the soil characteristics (edaphic properties) but upon its relation to the nature and history of landscapes. Today, there's an integration of the two disciplinary approaches as part of landscape and environmental sciences.

Pedologists are now also interested in the practical applications of a good understanding of pedogenesis processes (the evolution and functioning of soils), like interpreting its environmental history and predicting consequences of changes in land use, while agronomists understand that the cultivated soil is a complex medium, often resulting from several thousands of years of evolution. They understand that the current balance is fragile and that only a thorough knowledge of its history makes it possible to ensure its sustainable use.

Key Points

- Complexity in soil genesis is more common than simplicity.

- Soils lie at the interface of Earth's atmosphere, biosphere, hydrosphere and lithosphere. Therefore, a thorough understanding of soils requires some knowledge of meteorology, climatology, ecology, biology, hydrology, geomorphology, geology and many other earth sciences and natural sciences.

- Contemporary soils carry imprints of pedogenic processes that were active in the past, although in many cases these imprints are difficult to observe or quantify. Thus, knowledge of paleoecology, palaeogeography, glacial geology and paleoclimatology is important for the recognition and understanding of soil genesis and constitute a basis for predicting the future soil changes.

- Five major, external factors of formation (climate, organisms, relief, parent material and time), and several smaller, less identifiable ones, drive pedogenic processes and create soil patterns.

- Characteristics of soils and soil landscapes, e.g., the number, sizes, shapes and arrangements of soil bodies, each of which is characterized on the basis of soil horizons, degree of internal homogeneity, slope, aspect, landscape position, age and other properties and relationships, can be observed and measured.

- Distinctive bioclimatic regimes or combinations of pedogenic processes produce distinctive soils. Thus, distinctive, observable morphological features, e.g., illuvial clay accumulation in B horizons, are produced by certain combinations of pedogenic processes operative over varying periods of time.

- Pedogenic (soil-forming) processes act to both create and destroy order (anisotropy) within soils; these processes can proceed simultaneously. The resulting soil profile reflects the balance of these processes, present and past.

- The geological Principle of Uniformitarianism applies to soils, i.e., pedogenic processes active in soils today have been operating for long periods of time, back to the time of appearance of organisms on the land surface. These processes do, however, have varying degrees of expression and intensity over space and time.

- A succession of different soils may have developed, eroded and/or regressed at any particular site, as soil genetic factors and site factors, e.g., vegetation, sedimentation, geomorphology, change.

- There are very few old soils (in a geological sense) because they can be destroyed or buried by geological events, or modified by shifts in climate by virtue of their vulnerable position at the surface of the earth. Little of the soil continuum dates back beyond the Tertiary period and most soils and land surfaces are no older than the Pleistocene Epoch. However, preserved/lithified soils (paleosols) are an almost ubiquitous feature in terrestrial (land-based) environments throughout most of geologic time. Since they record evidence of ancient climate change, they present immense utility in understanding climate evolution throughout geologic history.

- Knowledge and understanding of the genesis of a soil is important in its classification and mapping.

- Soil classification systems cannot be based entirely on perceptions of genesis, however, because genetic processes are seldom observed and because pedogenic processes change over time.

- Knowledge of soil genesis is imperative and basic to soil use and management. Human influence on, or adjustment to, the factors and processes of soil formation can be best controlled and planned using knowledge about soil genesis.

- Soils are natural clay factories (clay includes clay mineral structures and particles less than 2 μm in diameter). Shales worldwide are, to a considerable extent, simply soil clays that have been formed in the pedosphere and eroded and deposited in the ocean basins, to become lithified at a later date.

Pedogenesis

Pedogenesis can be defined as the process of soil development. Late in the 19th century, scientists Hilgard in the United States and the Russian Dukuchaev both suggested independently that pedogenesis was principally controlled by climate and vegetation. This idea was based on the observation that comparable soils developed in spatially separate areas when their climate and vegetation were similar. In the 1940s, Hans Jenny extended these ideas based on the observations of many subsequent studies examining the processes involved in the formation of soils. Jenny believed that the kinds of soils that develop in a particular area are largely determined by five interrelated factors: climate; living organisms; parent material; topography; and time (Figure).

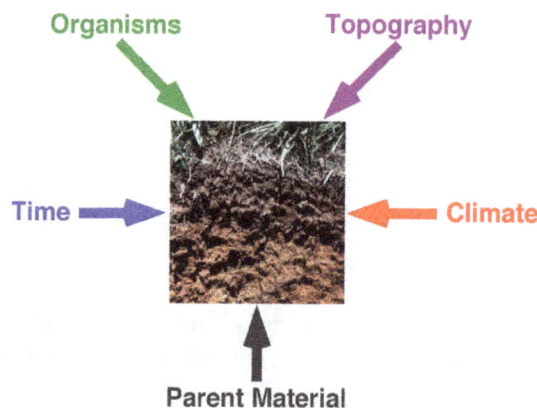

Figure The development of a soil is influenced by five interrelated factors: organisms, topography, time, parent material, and climate.

Climate plays a very important role in the genesis of a soil. On the global scale, there is an obvious correlation between major soil types and the Koppen climatic classification systems major climatic types. At regional and local scales, climate becomes less important in soil formation. Instead, pedogenesis is more influenced by factors like parent material, topography, vegetation, and time. The two most important climatic variables influencing soil formation are temperature and moisture. Temperature has a direct influence on the weathering of bedrock to produce mineral particles. Rates of bedrock weathering generally increase with higher temperatures. Temperature also influences the activity of soil microorganisms, the frequency and magnitude of soil chemical reactions, and the rate of plant growth. Moisture levels in most soils are primarily controlled by the addition of water via precipitation minus the losses due to evapotranspiration. If additions of water from precipitation surpass losses from evapotranspiration, moisture levels in a soil tend to be high. If the water loss due to evapotranspiration exceeds inputs from precipitation, moisture levels in a soil tend to be low. High moisture availability in a soil promotes the weathering of bedrock and sediments, chemical reactions, and plant growth. The availability of moisture also has an influence on soil pH and the decomposition of organic matter.

Living Organisms have a role in a number of processes involved in pedogenesis including organic matter accumulation, profile mixing, and biogeochemical nutrient cycling. Under equilibrium conditions, vegetation and soil are closely linked with each other through nutrient cycling. The cycling of nitrogen and carbon in soils is almost completely controlled by the presence of animals and plants. Through litter fall and the process of decomposition, organisms add humus and nutrients to the soil which influences soil structure and fertility. Surface vegetation also protects the upper layers of a soil from erosion by way of binding the soils surface and reducing the speed of moving wind and water across the ground surface.

Parent Material refers to the rock and mineral materials from which the soils develop. These materials can be derived from residual sediment due to the weathering of bedrock or from sediment transported into an area by way of the erosive forces of wind, water, or ice. Pedogenesis is often faster on transported sediments because the weathering of parent material usually takes a long period of time. The influence of parent material on pedogenesis is usually related to soil texture, soil chemistry, and nutrient cycling.

Topography generally modifies the development of soil on a local or regional scale. Pedogenesis is primarily influenced by topography's effect on microclimate and drainage. Soils developing on moderate to gentle slopes are often better drained than soils found at the bottom of valleys. Good drainage enhances an number of pedogenic processes of illuviation and eluviation that are responsible for the development of soil horizons. Under conditions of poor drainage, soils tend to be immature. Steep topographic gradients inhibit the development of soils because of erosion. Erosion can retard the development through the continued removal of surface sediments. Soil microclimate is also influenced by topography. In the Northern Hemisphere, south facing slopes tend to be warmer and drier than north facing slopes. This difference results in the soils of the two areas being different in terms of depth, texture, biological activity, and soil profile development.

Time influences the temporal consequences of all of the factors described above. Many soil processes become steady state overtime when a soil reaches maturity. Pedogenic processes in young soils are usually under active modification through negative and positive feedback mechanisms in attempt to achieve equilibrium.

Principal Pedogenic Processes

A large number of processes are responsible for the formation of soils. This fact is evident by the large number of different types of soils that have been classified by soil scientists. However, at the macro-scale we can suggest that there are five main principal pedogenic processes acting on soils. These processes are laterization, podzolization, calcification, salinization, and gleization.

Laterization is a pedogenic process common to soils found in tropical and subtropical environments. High temperatures and heavy precipitation result in the rapid weathering of rocks and minerals. Movements of large amounts of water through the soil cause eluviation and leaching to occur. Almost all of the byproducts of weathering, very simple small compounds or nutrient ions, are trans located out of the soil profile by leaching if not taken up by plants for nutrition. The two exceptions to this process are iron and aluminum compounds. Iron oxides give tropical soils their unique reddish coloring. Heavy leaching also causes these soils to have an acidic pH because of the net loss of base cations.

Podzolization is associated with humid cold mid-latitude climates and coniferous vegetation. Decomposition of coniferous litter and heavy summer precipitation create a soil solution that is strongly acidic. This acidic soil solution enhances the processes of eluviation and leaching causing the removal of soluble base cations and aluminum and iron compounds from the A horizon. This process creates a sub-layer in the A horizon that is white to gray in color and composed of silica sand.

Calcification occurs when evapotranspiration exceeds precipitation causing the upward movement of dissolved alkaline salts from the groundwater. At the same time, the movement of rain water causes a downward movement of the salts. The net result is the deposition of the trans located cations in the B horizon. In some cases, these deposits can form a hard layer called caliche. The most common substance involved in this process is calcium carbonate. Calcification is common in the prairie grasslands.

Salinization is a process that functions in the similar way to calcification. It differs from calcification in that the salt deposits occur at or very near the soil surface. Salinization also takes place in much drier climates.

Gleization is a pedogenic process associated with poor drainage. This process involves the accumulations of organic matter in the upper layers of the soil. In lower horizons, mineral layers are stained blue-gray because of the chemical reduction of iron.

Soil Morphology

Soil morphology is defined as the branch of soil science that deals with the description, using standard terminology, of in situ spatial organization and physical properties of soil regardless of potential use. Precise descriptions, using conventional terms, are necessary to all of the areas of science included in the field of soil science. The basic descriptors used today have been developed over the past 50 years and are continually added to. Prior to utilization of standard terminology, soils were described as clayey, sandy, stony, sedimentary, saline, marshy, dry or moist, heavy or light, soft or compact, fatty, friable, or lean. Soil horizons were described by terms such as 'gray watery sand' or 'rusty brown clay.' While these terms illustrate a fundamental understanding of soil properties, they do not impart any specific knowledge about the soils in question and cannot be compared with other soils described by other scientists. An objective, complete description of the soil is essential, because it serves as a basis for soil identification, classification, correlation, mapping, and interpretation.

Micromorphology

While soil micromorphology begins in the field with the routine and careful use of a 10x hand lens, much more can be described by careful description of thin sections made of the soil with the aid of a petrographic polarizing light microscope. The soil can be impregnated with an epoxy resin, but more commonly with a polyester resin (crystic 17449) and sliced and ground to 0.03 millimeter thickness and examined by passing light through the thin soil plasma.

Porosity

Porosity of topsoil is a measure of the pore space in soil which typically decreases as grain size increases. This is due to soil aggregate formation in finer textured surface soils when subject to soil biological processes. Aggregation involves particulate adhesion and higher resistance to compaction. Typical bulk density of sandy soil is between 1.5 and 1.7 g/cm^3. This calculates to a porosity between 0.43 and 0.36. Typical bulk density of clay soil is between 1.1 and 1.3 g/cm^3. This calculates to a porosity between 0.58 and 0.51. This seems counterintuitive because clay soils are termed heavy, implying lower porosity. Heavy apparently refers to a gravitational moisture content effect in combination with terminology that harkens back to the relative force required to pull a tillage implement through the clayey soil at field moisture content as compared to sand.

Porosity of subsurface soil is lower than in surface soil due to compaction by gravity. Porosity of 0.20 is considered normal for unsorted gravel size material at depths below the biomantle. Porosity in finer material below the aggregating influence of pedogenesis can be expected to approximate this value.

Soil porosity is complex. Traditional models regard porosity as continuous. This fails to account for anomalous features and produces only approximate results. Furthermore it cannot help model the influence of environmental factors which affect pore geometry. A number of more complex models have been proposed, including fractals, bubble theory, cracking theory, Boolean grain process, packed sphere, and numerous other models.

Active Layer

The active layer is that portion of the soil above permafrost that thaws and freezes seasonally. It plays an important role in cold regions because most ecological, hydrological, biogeochemical and pedogenic activity takes place within it. Changes in active layer thickness are influenced by many factors, including surface temperature, physical and thermal properties of the surface cover and substrate, vegetation, soil moisture and duration and thickness of snow cover. The inter-annual and spatial variations in thaw depth at point locations can be large, an artefact of year-to-year and micro topographic variations in both surface temperature and soil moisture, and so presents monitoring challenges. When the other conditions remain constant, changes in active layer thickness could be expected to increase in response to climate warming, especially in summer.

Steps Showing the Thawing and Freezing Cycle

Under the tundra in the northern part of the Arctic Refuge, the active layer begins to thaw after the winter snows melt in early summer. Snow can fall in any month, but it doesn't usually stay on the ground until September. What date do you think the active layer is melted deepest?

Here's how the active layer changes throughout the year (These dates are averages. The actual dates on any year may be slightly earlier or later.):

On May 15, the tundra soil is still buried under winter snow, and the active layer is completely frozen.

By June 1, the snow has begun to melt, but the active layer remains frozen.

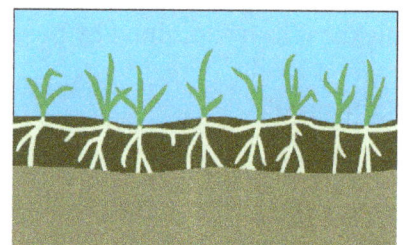

On June 15, the active layer is already thawed half way (50%) to its maximum depth. It is about 9 inches (23 cm) deep. The sun remains above the horizon for 24 hours each day, and plants are producing new growth.

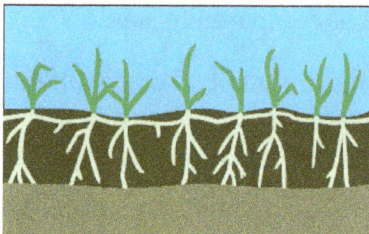

By July 1, the active layer has thawed to 75% of its total depth. It is about 13.5 inches (34.5 cm) deep.

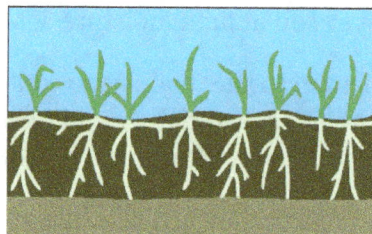

On July 15, the active layer has thawed to 85% of its total depth. It is about 15 inches (38 cm) deep.

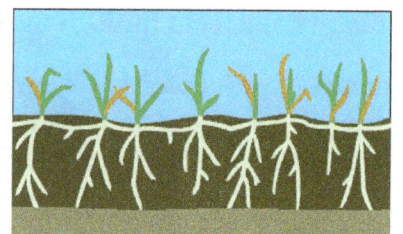

On August 1, the active layer has thawed to 90% of its total depth. It is about 16 inches (40.5 cm) deep.

By August 15, the active layer has thawed to 92% of its total depth. It is about 16.5 inches (42 cm) deep. Autumn has come to the tundra. The air is cooler, and plants are finishing their growth for the year.

On September 1, the active layer has thawed to 94% of its total depth. It is about 16.75 inches (42.5 cm) deep. Air temperatures are now below freezing, and the soil surface begins to refreeze.

On September 15, the active layer has thawed to 96% of its total depth. It is about 17.25 inches (43.25 cm) deep. The soil surface continues to freeze down into the active layer.

By October 1, the active layer has thawed to 98% of its total depth. It is about 17.75 inches (44.5 cm) deep.

On October 15, the active layer has thawed to its maximum depth of 18 inches (about 45.5 cm). The soil above it is refreezing rapidly. Did you guess that the active layer would reach its maximum depth in mid-October?

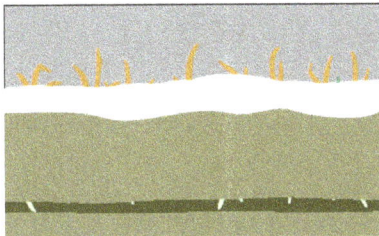

By November 1, there is just a narrow region above the base of the active layer that is not frozen.

By November 15, the active layer has completely frozen, and will remain this way until early summer.

Soil Map

Purpose of Soil Maps

A map does not represent the exact reality of the land charted, but instead provides a structured representation of knowledge about the distribution of soils across the landscape. It is an approximation of the reality of the land, and one whose accuracy increases at higher densities of observation and more detailed scales. Soil maps of a given area constitute a visualization of the intrinsic properties of the soils in question and a delimitation of their specific areas of distribution. The division of a given landscape into supposedly homogenous areas constitutes a modeling exercise which represents the distribution of soils in the landscape. Mapping this information involves a considerable effort of synthesis and concretion. A soil map should harmoniously combine the working scale, the amount of information that needs to be represented, the quality of definition and unit of delimitation and the interpretative capacity of those who use this information.

MAPA DE SÒLS 1:100.000 DE L'ÀREA REGABLE
PEL CANAL D'ARAGÓ I CATALUNYA

Map Provided by the DAR

The Purposes of Soil Maps are therefore to:

- Synthesize the key properties of the soils in a single document using a set of codifications and legends that allow a rapid interpretation of the results.

- Describe the distribution of soil units, grouped under more or less generalized or detailed concepts, at a suitable working scale.

- Provide a graphic document which synthesizes the inventory and/or evaluation of the soil units and their corresponding distribution over the land. con la distribución correspondiente encima del territorio.

- Contribute to the dissemination of knowledge about soils and their spatial distribution and properties. The soil map is often the only document allows a non-specialist access to soil information.

- Allow rapid access to soil information. As a specific, visual, synthetic and bi-dimensional representation of the soils of a specific area, the soil map should facilitate access to information which will enable rational use of the soil data provided.

References

- Buol, Stanley W.; Southard, Randal J.; Graham, Robert C.; McDaniel, Paul A. (2003). Soil Genesis and Classification, 5th Edition. Ames, Iowa: Iowa State Press, A Blackwell Pub. Co. p. 494. ISBN 0-8138-2873-2

- Pedology, science: britannica.com, Retrieved 26 June 2018

- Soil Survey Staff (1993). Soil Survey Manual. Washington D.C.: U. S. Government Printing Office. Soil Conservation Service, United States Department of Agriculture Handbook 18. Archived from the original on 2007-02-07. Retrieved 2006-11-03

- What-is-pedology: academicroom.com, Retrieved 20 April 2018

- Macphail, Richard I; Courty, Marie-Agnès; Goldberg, Paul (January 1990). "Soil micromorphology in archaeology". Endeavour. 14 (4): 163–171. doi:10.1016/0160-9327(90)90039-t. ISSN 0160-9327

- Soil-morphology, agricultural-and-biological-sciences: sciencedirect.com, Retrieved 12 March 2018

Chapter 4

Edaphology

Edaphology is a branch of soil science concerned with the study of the influence of soils on plants and other living organisms. This chapter has been carefully written to provide an easy understanding of the varied facets of edaphology, such as agricultural soil science, environmental soil science, bioeffector, plant nutrition, phosphate solubilizing bacteria, etc. for a holistic understanding of edaphology.

Edaphology is one of two soil sciences which also consist of Pedology. Specifically, edaphology is the study in how soils influence and interact on/with living things, especially those of plants. It is the study of the ecological relationship soil has with land cultivation practices and plants. For the cultivation of land, edaphology also focuses on the conservation of soils and possible loss or erosion due to certain practices. Viticulturists often rely on edaphologists to assist and advise on the initial planting and planning of a vineyard including vine species selection for the type of soil available.

In the study of soil science, soil is considered to have three dimensions; breadth, length and depth. Edaphology further studies the composition of soil which makes up its overall volume; mineral mater, organic matter, soil water and soil air. It also makes note that soil is comparable to the systems of animals essential to its life:

- Respiratory system - how the air circulates and exchanges gases
- Digestive system - the decomposition of organic matter
- Circulatory system - how water moves in the soil
- Excretory system - the leaching out of excess salt
- The brain - composition of clay
- Color
- Height - the soil's depth

Within viticulture, the study of soil and how it co-exists with organic life is essential to growing prosperous grape crops and staying ahead of evolutionary vine changes and climate changes.

Agricultural Soil Science

Agricultural soils are managed systems, to serve human purposes. First purpose is providing the populations with food. Second is the production of plants of high quality, as well as animal feed with sufficient protein content. Furthermore, from the economic point of view, agriculture needs yields providing sufficient income. The essential plant nutrient N has to be used appropriately for these several goals. In a longer time appreciation, soil fertility is also to be considered, soil organic matter being the main reserve of plant nutrients. Providing growing plants with enough and especially with adequate amounts of N needs (i) the knowledge about the demand of the special crop, (ii) considerations on the organic N pool in soil and (iii) on its transformation in plant-available inorganic forms (NH_4^+ and NO_3^-) as well as (iv) taking in balance the potential biological N_2 fixation and the N deposition. The mineralization of N from the organic pool is provided by soil microorganisms, using soil organic matter as source of energy and nutrient. Weather conditions strongly affect the N transformations in soil. Because weather conditions are not predictable with enough precision, the calculation of N flow from the immobile into the plant-available form is generally a site-specific determination relying on the average weather data from the last few years. As an almost general rule, NO_3^- accumulates periodically in soils. However, the main accumulation of NO_3^- occurs (i) after application of certain fertilizers or manure, (ii) after harvest, when plant residues are degraded by soil microorganisms and (iii) in late winter before plant growth. When no growing plants take up the NO_3^-, it may be denitrified or water transferred to deeper soil parts and to groundwater or surface water.

Environmental Soil Science

Environmental soil science is the study of the interaction of humans with the pedosphere as well as critical aspects of the biosphere, the lithosphere, the hydrosphere, and the atmosphere. Environmental soil science addresses both the fundamental and applied aspects of the field including: buffers and surface water quality, vadose zone functions, septic drain field site assessment and function, land treatment of wastewater, storm water, erosion control, soil contamination with metals and pesticides, remediation of contaminated soils, restoration of wetlands, soil degradation, nutrient management, movement of viruses and bacteria in soils and waters, bioremediation, application of molecular biology and genetic engineering to development of soil microbes that can degrade hazardous pollutants, land use, global warming, acid rain, and the study of anthropogenic soils, such as terra preta. Much of the research done in environmental soil science is produced through the use of models.

Bioeffector

The use of bioeffectors, formally known as plant biostimulants, has become common practice in agriculture and provides a number of benefits in stimulating growth and protecting against stress. A biostimulant is loosely defined as an organic material and/or microorganism that are applied to enhance nutrient uptake, stimulate growth, and enhance stress tolerance or crop quality.

Plant biostimulants, sometimes referred to as agricultural biostimulants, are a diverse classification of substances that can be added to the environment around a plant and have positive effects on plant growth and nutrition, but also on abiotic and biotic stress tolerance. Although most plant biostimulants are added to the rhizosphere to facilitate uptake of nutrients, many of these also have protective effects against environmental stress such as water deficit, soil salinization and exposure to sub-optimal growth temperatures. Biostimulants are not nutrients per se; instead they facilitate the uptake of nutrients or beneficially contribute to growth promotion or stress resistance. A newly emerged paradigm emphasizes that plants are not standalone entities within their environments; instead they are host and partner to microorganisms of bacteria and fungi; plants are a host to numerous micro biota and those associations, both outside and within its tissues, allow them to respond and adapt to abiotic and biotic stress. Reasonably, if we functionally optimize these associations, we may strengthen their role in plant stress protection.

The industry definition of biostimulants was originally proposed in 2012 and stated: "Plant biostimulants contain substance(s) and/or microorganisms whose function when applied to plants or the rhizosphere is to stimulate natural processes to enhance/benefit nutrient uptake, nutrient efficiency, tolerance to abiotic stress, and crop quality. Biostimulants have no direct action against pests, and therefore do not fall within the regulatory framework of pesticides". Biostimulants were loosely defined for a long time and often regarded dubiously because of their aggregate nature and the inherent difficulty to determine which specific components were making beneficial contributions. The definition proposed by du Jardin "A plant biostimulant is any substance or microorganism applied to plants with the aim to enhance nutrition efficiency, abiotic stress tolerance and/or crop quality traits, regardless of its nutrients content" represents the clearest and most concise way to define biostimulants.

Algal Extracts

Seaweed extracts (SWE) as biostimulants are emerging as commercial formulations for use as plant growth-promoting factors and a method to improve tolerance to salinity, heat, and drought. Algal extracts target a number of pathways to increase tolerance under stress. Seaweeds are red, green, and brown macro algae that represent 10% of marine productivity. Macro algae have been used as organic fertilizers for thousands of years and are still in use. Currently, there are over 47 companies producing and marketing various algal extracts for agricultural use; the majority of the formulations are from the brown algae, Ascophyllum nodosum.

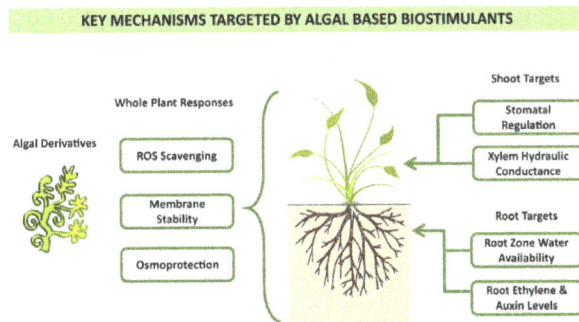

Fig. Summary of main key mechanisms targeted by algal-based biostimulants

While the growth-promoting effects of seaweed extracts have been documented in many species, very little is actually known about the mechanisms behind these effects. The variable and complex nature of these substances makes it difficult to determine exactly which components are playing a key role. Commercial formulations of SWEs are often proprietary, and the composition is largely dependent on the method of extraction. Indeed, characterization of the actual composition of most common algal-based commercial products would be useful first step to better hypothesize and/or depict a cause–effect relationship of their mechanism of action. Mechanical disruption, pulverization, acid or alkali extractions are some of the more common methods employed. Most commercial products are derived from red (ex Lithothamnium calcareum) and brown (ex Ascophyllum nodosum, Durvillaea potatorum) macro algae. The role of SWEs and cold tolerance is now emerging. Very recent work has focused on SWEs and their ability to enhance tolerance to chilling stress. When multiple extracts were tested for their ability to enhance cold tolerance in maize only extracts rich in Zn and Mn were able to enhance tolerance through enhanced ROS responses. In this case, the protective effects likely stem from supplying plants with micronutrients that play a role as co-factors in anti-oxidative enzymes. These results indicate that nutrient deficiency stress induced by cold can be overcome by supplying SWEs rich in micronutrients to improve oxidative stress tolerance. Previous studies with corn seedlings under root chilling stress supplemented with micronutrients demonstrated the utility of nutrient seed priming.

Some work has been done in model systems with the goal of determining the physiological and molecular responses induced by SWEs. In order to better understand the active components of A. nodosum, Rayirath et al. separated the organic-sub-fractions of extracts and tested them with Arabidopsis thaliana and freezing experiments. Plants grown in vitro with sub-fractions added to the substrate or in "Peat pellet freezing assays" irrigated with sub-fractions were tested for freezing tolerance. The authors found that the ethyl acetate extracted fraction, rich in fatty acids and sterols enhanced freezing tolerance over water treated (controls) at temperatures from −2.5 to −5.5 °C. Treated plants maintained faster rates of recovery, greater membrane integrity, and had 70% less chlorophyll damage upon freezing recovery as well as increased expression of key freezing tolerance genes such as RD29A, COR15A, and CBF3. Priming of key tolerance genes prior to exposure to stress greatly increases tolerance in many cases. The lipophilic components were found to be rich in fatty acids such as butyric acid, palmitic acid, oleic acid, linoleic acid the sterol fucosterol. These extracts increased proline content and total soluble sugars, contributing to freezing tolerance. A. nodosum extracts have even been used to reduce cold stress sensitivity in Kappaphycus alvarezii. Kappaphycus alvarezii is a red algae and the most important source of carrageenan; which are hydrophilic colloids largely used in foods and dairy products. Algal extracts have also been used on Kentucky bluegrass (Poa pratensis L. cv. Plush) to alleviate salinity stress from saline watering in turf grass experiments. Similarly SWE-based cytokines have been used on creeping bent grass (Agrostis stolonifera L.) to improve tolerance to heat stress. SWEs from A. nodosum have also been used for ornamental plants, such as Spiraea nipponica "Snowmound" and Pittosporum eugenioides "Variegatum", to enhance drought tolerance. Treated plants showed higher phenolic, proline, and flavonoid content while demonstrating improved physiology under mild drought stress conditions.

In horticultural crops and trees, SWE have been largely used for similar purposes. A. nodosum SWE increased RWC, Fresh Weight, and Dry Weight in spinach (Spinacia oleracea L.) plants under drought stress with some adverse effects on the nutritional value through reduced ferrous ion chelating ability. SWE applied to seedlings of lettuce (Lactuca sativa L.) enhanced cotyledon growth similar to fertilization with potassium.

Foliar application of marine bioactive substances (isopropanol extracts from microalgae) to grape plants (Vitis vinifera L) increased leaf water potential and stomatal conductance under drought stress. Consistent with an improved stomatal response, it was also observed that K^+ and Ca^{2+} fluxes at the stomatal level were higher in treated plants. Commercial formulations of A. nodosum have been tested on almond plants (Prunus dulcis [Mill.] D. A. Webb), which demonstrated increased growth and accumulation of K^+. In conditions with ample K^+ both MegaFol and GroZyme (Valagro, Atessa, Chieti, ITALY) increased leaf area and number of leaves greater than controls treated with water or K^+. In K^+-deficiency conditions only MegaFol and a foliar application of K^+ was able to stimulate growth, although at lower levels than observed with adequate K^+ nutrition. Accumulation of K+ is an essential step in protecting against both ionic and osmotic stress and may contribute to tolerance. Orange trees, Citrus sinensis L., subjected to drought stress and treated with commercial extracts of A. nodosum had better water relations and increased water use efficiency (WUE) under irrigation at 50% restitution of evapo-transpired water. The promise of biostimulants to increase drought tolerance and WUE holds great potential for drought prone regions where horticultural crops and fruit trees are agronomically important but water availability is becoming less reliable due to urbanization and climate change.

As earlier noted, almost all of the above-mentioned experiments with SWE use commercial formulations. This may be of some concern, due to the variable nature of these products and formulation methods. A recent transcriptomic study using A. thaliana plants treated with two different commercial A. nodosum extracts showed that not all extracts are alike. One commercial product resulted in dysregulation of 4.47% of the transcriptome while the other extract only affected 0.87%. Since transcriptional priming is likely a key component in enhancing abiotic stress tolerance using SWEs, these differences imply significant variability in responses elicited. Compositions of the extracts differed greatly, indicating that choice of commercial product may have a significant effect on plant responses. Commercial formulations are often proprietary and the exact composition and extraction methods, shifting the burden to the research community to analyze and isolate the active components in these products. In order to identify and characterize how these SWEs affect plants, some form of standardization is necessary.

Carbohydrates, Proteins, Amino Acids, and Lipids

Protein hydrolysates are mixtures of polypeptides, oligopeptides, and free amino acids derived from partial hydrolysis of agricultural by-products from animals and plants. Carbohydrates, proteins, amino acids, and lipids may increase stress tolerance through different (Fig.). The effects of amino acids on ion fluxes across membranes have been clearly established, with most having a positive effect on reducing NaCl-induced potassium efflux. Protein hydrolysates (PH) are often sold as formulations that include plant growth regulators. The bulk of PH products, over 90%, are produced from chemical hydrolysis of animal by-products while enzymatically processed plant-based products are a recent development.

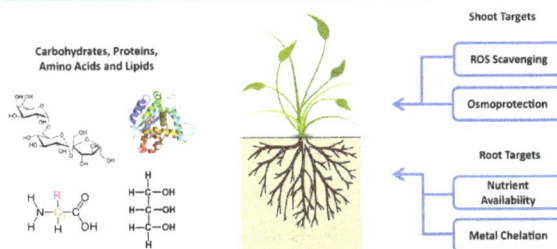

KEY MECHANISMS TARGETED BY CARBOHYDRATES, PROTEINS, AMINO ACIDS AND LIPIDS BASED BIOSTIMULANTS

Fig. Summary of main key mechanisms targeted by carbohydrate-, protein-,
amino acid-, and lipid-based biostimulants

Megafol (Valagro, Atessa, Chieti, ITALY) is a commercial biostimulant comprising vitamins, amino acids, proteins, and betaines from plant and algal extracts. Application of Megafol to tomato plants under drought stress enhanced induction of a number of drought responsive genes such as tomato orthologs of RAB18 and RD29B. Treated plants also had higher fresh weight and relative water content under drought stress, indicating a protective effect on water status and stress responsive genes. When hydrolysate-based biostimulants from alfalfa (Medicago sativa L.), containing triacontanol (TRIA) and indole-3-acetic acid (IAA), were applied to maize plants under salt stress, the protective effects were amplified. Treated plants had higher flavonoid, proline, and potassium content in salt stress conditions over untreated controls. Extracts that are rich in amino acids may play a role in increasing cold tolerance. When lettuce plants (Lactuca sativa) were treated with an amino acid mixture, derived from enzymatic hydrolysis of proteins, (Terra-Sorb) and subjected to cold, treated plants had higher fresh weights and improved stomatal conductance. Use of animal derived amino acid hydrolysates on strawberry plants after transplantation and cold stress did not improve survival though some growth promotions were observed in the absence of stress. Perennial Rye-grass (Lolium perenne L.) treated with hydrolyzed amino acids and high temperatures (36 °C) had improved photosynthetic efficiency over control plants.

Mutants of A. thaliana deficient in production of proline have stress sensitive phenotypes. These plants can have their phenotype rescued with exogenous application of l-proline, a common amino acid available in biostimulant formulations of various amino acids and hydrolysate mixtures. Hydrolysates from wheat germs show strong anti-oxidant and free radical scavenging properties as well as the ability to chelate some metals.

Lettuce (Lactuca sativa L.) is particularly salt sensitive and the addition of plant-derived protein hydrolysates improved fresh yield, dry biomass, and root dry weight as well as increased concentrations of osmoyltes, glucosinolates and the composition of sterols and terpenes. Hydrolysates have applications for trees, which require considerable investment costs and can be vulnerable to drought. Japanese persimmon trees, Diospyros kaki L. cv. "Rojo Brillante" grafted on Diospyros lotus L., are particularly sensitive to drought stress. Treatment of these trees with calcium protein hydrolysates decreased chloride uptake under saline irrigation, lowered water potentials as well as increased the concentration of compatible solutes, all of which would enhance plant growth under saline stress.

Recent reports indicate that melatonin, derived from l-tryptophan via the shikimate pathway, can prime seeds to tolerate adverse environmental conditions at imbibition and germination stages. Corn seeds pre-treated with melatonin show increased tolerance to chilling stress upon germination, indicating a priming effect by melatonin. Melatonin may prove to be an effective biostimulant for improving stress tolerance of seedlings.

Glycinebetaine is a compatible solute accumulated in many plants in response to salt stress. Exogenous application of glycinebetaine has increased tolerance for environmental stresses such as drought, chilling, freezing, salinity, and oxidative stress. Foliar application of glycinebetaine results in rapid uptake by leaves and concentration in meristematic tissues. Rapid uptake and localization of glycinebetaine in these most vulnerable tissues are particularly beneficial in chilling and freezing stress where glycinebetaine can exert a protective effect. Transgenic plants of various species expressing two biosynthetic genes, codA and betA, produce more glycinebetaine and had an increased tolerance to abiotic stress. Exogenous application of small amounts of compatible solutes such as proline and betaine to barley roots resulted in an immediate reduction of NaCl-induced efflux of K+, indicating that ion fluxes across the membrane can be affected by relatively low concentrations of compatible solutes. The cause–effect relationship between accumulation of compatible solutes and stress protection still remains to be fully understood. However, a better understanding of the specific mechanisms of action of these molecules is becoming increasingly important if we want to make predictions on which combination of biostimulants can be more effective.

Humic and Fulvic Acids

Humic and fulvic substances are the major organic components of lignites, soil, and peat. Humic and fulvic acids are produced by the biodegradation of organic matter resulting in a mixture of acids containing phenolate and carboxyl groups. Fulvic acids are humic acids with a higher oxygen content and lower molecular weight. A number of examples exist indicating the potential for these substances to improve abiotic stress tolerance in plants (Fig.). Pre-treatment of tall fescue (Festuca arundinacea Schreb.) and creeping bent grass (Agrostis palustris Huds. A.) with seaweed extract and humic acid increased leaf hydration under dry soil conditions as well as root growth, shoot growth, and antioxidant capacity. Further studies with bent grass showed these extracts, high in cytokinins, combined with humic acid increased drought tolerance as well as endogenous cytokinin content.

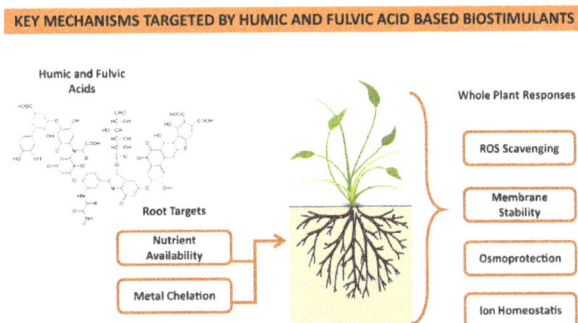

Fig. Summary of main key mechanisms targeted by humic- and fulvic acid-based biostimulants

Treatment of bell pepper (Capsicum annuum L. cv. Demre) with humic acid and phosphorous resulted in plants with reduced Na content and elevated N, P, K, Ca, Fe, Mg, S, Mn, and Cu ion contents in roots and shoots, which were associated with a general protective effect under mild salinity stress. Application of humic acids to common bean (Phaseolus vulgaris L.) under high salinity (120 mM NaCl) increased endogenous proline levels and reduced membrane leakage, which are both indicators of better adaptation to saline environments.

Humic acid extracts seem to be beneficial also for field crop monocots. Extracts from vermicompost applied to rice (Oryza sativa L.) played a role in activating anti-oxidative enzymatic function and increased ROS scavenging enzymes. These enzymes are required to inactivate toxic-free oxygen radicals produced in plants under drought and saline stress. One possible mode of action for vermicompost may be the differential regulation of proton ATPases located in the vacuolar and plasma membranes. When Micro-Tom tomato plants were treated with vermicompost, plasma membrane proton extrusion was increased by over 40% which facilitated acid growth and nutrient uptake potential. Interestingly, the auxin insensitive mutant diageotropica (dgt) showed no increase in proton extrusion, indicating that humic substance may increase root growth through mediating auxin signalling.

Microorganisms Affecting Stress Tolerance

While plants are known to establish symbiotic relationships with bacteria, our understanding of those relationships under abiotic stress is rudimentary. However, some of the targets of microorganisms that increase abiotic stress tolerances have been identified (Fig.). Bacteria with the potential to act as biostimulants have been isolated from a number of ecosystems with saline, alkaline, acidic, and arid soils. These bacteria belong to several genera such as Rhizobium, Bradyrhizobium, Azotobacter, Azospirillum, Pseudomonas, and Bacillus. Members of these genera have developed strategies to adapt and thrive under adverse conditions. Amongst these adaptations, alterations to the composition of the cell wall and the ability to accumulate high concentrations of soluble solutes are common. These allow for enhanced water retention and increased tolerance to osmotic and ionic stress. Cell wall composition is altered through enrichment for exopolysaccharides (EPS) and lipopolysaccharide–proteins and polysaccharide–lipids which my form a protective biofilm on the root surface. Plant growth-promoting rhizobacteria (PGPR) inoculated soils can ameliorate plant abiotic stress responses. A number of recent reviews have extensively covered the protective effects of Rhizobium against abiotic stress in plants. Most documented growth enhancement determined by these bacteria is associated with high level of IAA, which has been proven to alleviate salt stress and EPS production that may help in maintaining a film of hydration around the roots and/or help re-establishing favorable water potential gradients under water limitations. These functions have been proven useful under saline stress, extremes of temperature, pH, salinity, and drought. Inoculation of maize with Azotobacter strains has been shown to have general positive effects under saline stress by facilitating uptake of K+ and exclusion of Na+ as well as increasing phosphorous and nitrogen availability. In wheat, inoculation of salt tolerance Azobacter strains increased biomass, nitrogen content, and grain yield under salt stress.

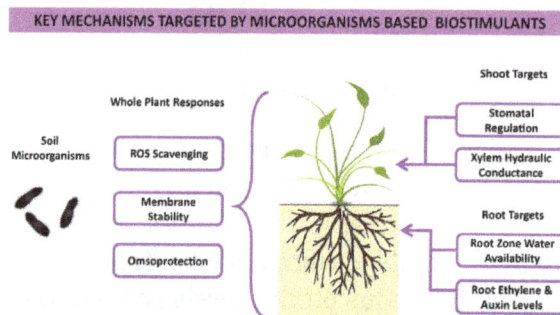

Fig. Summary of main key mechanisms targeted by microorganism-based biostimulants

Tolerance to salt stress varies within these microorganisms and their tolerance can confer advantages to the host relationship under stress conditions. When two legumes, pea (Pisum sativum) and fava bean (Vicia faba), were inoculated with two different strains of Rhizobium leguminosarum, a salt-tolerant (GRA19) and salt-sensitive (GRL19) plants inoculated with the salt-tolerant strain performed better under moderate salt stress. The authors further found that pea plants had larger nodules and high levels of nitrogen fixation under salt stress when inoculated with GRA19, the salt-tolerant strain of R. leguminosarum. Similar results have been observed for non-symbiotic free-living soil bacteria that are capable of fixing nitrogen. Azospirillum brasilense is closely associated with the plant rhizosphere and can colonize the surface of roots. When chickpea (Cicer arietinum L.) and faba bean were inoculated with A. brasilense, they experienced enhanced nodulation by native rhizobia and greater tolerance to salt stress. Another free-living nitrogen-fixing species, Azotobacter chrococcum A2 demonstrated salt tolerance. Inoculation with A. chrococcum has been shown to increase yields of pea, potato, rice, wheat, and cotton in saline-arid soils. Increased root length and shoot growth was also observed with inoculation with significant positive yield effects for wheat (from 2.8 to 3.5 t ha^{-1} when grown in conjunction with A. chrococcum).

In barley, Hartmannibacter diazotrophicus E19 (T) is capable of colonizing roots in saline conditions. Inoculation of roots in saline soil increased root and shoot mass significantly, 308 and 189%, respectively. Inoculated roots also had increased relative water content over three and a half times that of control plants. High concentrations of salt can also be inhibitory to rhizobial bacteria. While certain strains of R. leguminosarum, such as viciae SAAN1, are very salt tolerant and able to withstand up to 0.34 M NaCl, they often show lower rates of nodulation in saline soils. These strains are often less competitive with natural rhizobial populations, however.

The stress protection of bacterial biostimulants to rain fed field crops can be of particular relevance under increasing temperatures foreseen by most climate change prediction models. Wheat inoculated with the thermo tolerant Pseudomonas putida strain AKMP7 significantly increased heat tolerance. Inoculated plants had increased biomass, shoot and root length, and seed size. ROS generation under stress treatment was also lessened, with lower levels of expression observed for ROS response genes such as superoxide dismutase, ascorbate peroxidase, and catalase. Similar results have been found with sorghum and other Pseudomonas putida AKMP strains. Psychrophilic (cold-adapted) microorganisms are capable of surviving in extreme conditions and their interactions with plants provide potential mechanisms for improving tolerance. While many strains of soil bacteria with growth-promoting properties have been isolated from low-temperature conditions, few have been tested in conjunction with plants subjected to cold stress.

Wheat inoculated with the cold-tolerant plant growth-promoting bacteria Pantoea dispersa showed improved growth and nutrient uptake, likely due to the solublization of phosphorous and production of IAA.

Inoculation of soil with psychrotolerant (cold tolerant) bacteria can play a role in chilling tolerance. The psychrotolerant soil bacterium, Burkholderia phytofirman, is a plant-growth-promoting rhizobacterium (PGPR) that is capable of colonizing multiple plant species. B. phytofirman was shown to play a role in enhancing chilling tolerance in Vitis vinifera L. by increasing ROS scavenging metabolites and stress-induced genes. Inoculated plants also recovered faster from chilling stress, returning to normal metabolic levels more quickly than controls. B. phytofirman inoculation also alters carbohydrate metabolism and accumulation while having a protective effect on net photosynthesis during cold acclimation and stress.

Tomato plants (Solanum lycopersicum cv Mill) were inoculated with cold-tolerant strains of Pseudomonas vancouverensis, and frederiksbergensis as well as Flavobacterium glaciei that were isolated from agricultural fields during winter. Treated tomato seedlings were subjected to a week of chilling stress at 15C and inoculation three of these strains showed significantly reduced electrolyte leakage and ROS activity. Improved stress tolerance and growth-promoting effects of microorganism treatments have been seen in other species also. Inoculation of lettuce (Lactuca sativa L., cv Mantecosa) seeds with A. brasilense increased germination in the presence of salt and demonstrated tolerance through higher total fresh and dry weights of plants at harvest. Additional experiments studying these effects have shown increased biomass, chlorophyll, ascorbic acid content, antioxidant content, and post-harvest shelf life after being subjected to salt stress. Sweet pepper (Capsicum annuum L.) inoculated with A. brasilense and Pantoea dispersa was not affected by moderate levels of salinization, up to 80 mM NaCl, while uninoculated control plants demonstrated lower DW starting at 40 mM NaCl.

Triticum aestivum cv. Buck Ombú inoculated with A. brasilense sp. 245 and subjected to salt stress (320 mM NaCl) and osmotic stress (20 and 30% PEG 6000) had higher FW, DW, and RWC than non-inoculated controls. Analysis of phospholipids and fatty acid composition in inoculated wheat indicated that the distribution profiles of major root phospholipids are altered in inoculated plants, possibly contributing to the increased tolerance. Wheat inoculated with Azospirillum lipoferum and irrigated with 80 mM NaCl had significantly higher leaf and root dry weight than uninoculated controls.

While the mechanisms by which A. brasilense confer tolerance to osmotic stress are not clear, some evidence indicates that inoculation induces wider xylem vessels and greater hydraulic conductance. In inoculated tomato plants subjected to water stress similar changes have been observed, such as larger xylem vessel area, higher stem-specific hydraulic conductivity, and thicker stems. Pepper plants co-inoculated with A. brasilense and Pantoea dispersa accumulated more dry matter under salt stress. Inoculated plants showed higher stomatal conductance and rates of photosynthesis under salt stress. The chlorophyll concentration and efficiency of photosystem II were not affected in inoculated plants under stress conditions.

Inhibition of root growth under salt stress conditions is well documented. One of the primary causes of this inhibition is the production and perception of ethylene in the roots. Plants and PGPR both have ACC-deaminases, which possess the ability to lower the concentration of ethylene in the roots and root zone. PGPR-derived ACC-deaminases can reduce ethylene induced inhibition by reducing root zone ethylene and contribute to maintain relatively higher root-to-shoot ration, a trait that would result beneficial under water shortage.

Biostimulant treatments of agricultural crops have the potential to improve plant resilience to environmental perturbations. In order to fine-tune application rates, biostimulant-plant specificities and techniques is identified that may yield highest impact on stress protection; high priority should be given to better understanding of the causal/functional mechanism of biostimulants.

Plant Nutrition

Plants use inorganic minerals for nutrition, whether grown in the field or in a container. Complex interactions involving weathering of rock minerals, decaying organic matter, animals, and microbes take place to form inorganic minerals in soil. Roots absorb mineral nutrients as ions in soil water. Many factors influence nutrient uptake for plants. Ions can be readily available to roots or could be "tied up" by other elements or the soil itself. Soil too high in pH (alkaline) or too low (acid) makes minerals unavailable to plants.

Fertility or Nutrition

The term "fertility" refers to the inherent capacity of a soil to supply nutrients to plants in adequate amounts and in suitable proportions. The term "nutrition" refers to the interrelated steps by which a living organism assimilates food and uses it for growth and replacement of tissue. Previously, plant growth was thought of in terms of soil fertility or how much fertilizer should be added to increase soil levels of mineral elements. Most fertilizers were formulated to account for deficiencies of mineral elements in the soil. The use of soilless mixes and increased research in nutrient cultures and hydroponics as well as advances in plant tissue analysis have led to a broader understanding of plant nutrition. Plant nutrition is a term that takes into account the interrelationships of mineral elements in the soil or soilless solution as well as their role in plant growth. This interrelationship involves a complex balance of mineral elements essential and beneficial for optimum plant growth.

Essential Versus Beneficial

The term essential mineral element (or mineral nutrient) was proposed by Arnon and Stout (1939). They concluded three criteria must be met for an element to be considered essential. These criteria are: 1. A plant must be unable to complete its life cycle in the absence of the mineral element. 2. The function of the element must not be replaceable by another mineral element. 3. The element must be directly involved in plant metabolism. These criteria are important guidelines for plant nutrition but exclude beneficial mineral elements. Beneficial elements are those that can compensate for toxic effects of other elements or may replace mineral nutrients in some other less specific function such as the maintenance of osmotic pressure. The omission of beneficial nutrients in commercial production could mean that plants are not being grown to their optimum genetic potential but are merely produced at a subsistence level. This discussion of plant nutrition includes both the essential and beneficial mineral elements.

The Mineral Elements

There are actually 20 mineral elements necessary or beneficial for plant growth. Carbon (C), hydrogen (H), and oxygen (O) are supplied by air and water. The six macronutrients, nitrogen (N), phosphorus (P), potassium (K), calcium (Ca), magnesium (Mg), and sulfur (S) are required by plants in large amounts. The rest of the elements are required in trace amounts (micronutrients). Essential trace elements include boron (B), chlorine (Cl), copper (Cu), iron (Fe), manganese (Mn), sodium (Na), zinc (Zn), molybdenum (Mo), and nickel (Ni). Beneficial mineral elements include silicon (Si) and cobalt (Co). The beneficial elements have not been deemed essential for all plants but may be essential for some. The distinction between beneficial and essential is often difficult in the case of some trace elements. Cobalt for instance is essential for nitrogen fixation in legumes. It may also inhibit ethylene formation (Samimy, 1978) and extend the life of cut roses (Venkatarayappa et al., 1980). Silicon, deposited in cell walls, has been found to improve heat and drought tolerance and increase resistance to insects and fungal infections. Silicon, acting as a beneficial element, can help compensate for toxic levels of manganese, iron, phosphorus and aluminum as well as zinc deficiency. A more holistic approach to plant nutrition would not be limited to nutrients essential to survival but would include mineral elements at levels beneficial for optimum growth. With developments in analytical chemistry and the ability to eliminate contaminants in nutrient cultures, the list of essential elements may well increase in the future.

The Mineral Elements in Plant Production

The use of soil for greenhouse production before the 1960s was common. Today a few growers still use soil in their mixes. The bulk of production is in soilless mixes. Soilless mixes must provide support, aeration, nutrient and moisture retention just as soils do, but the addition of fertilizers or nutrients are different. Many soilless mixes have calcium, magnesium, phosphorus, sulfur, nitrogen, potassium and some micronutrients incorporated as a pre-plant fertilizer. Nitrogen and potassium still must be applied to the crop during production. Difficulty in blending a homogenous mix using pre-plant fertilizers may often result in uneven crops and possible toxic or deficient levels of nutrients. Soilless mixes that require addition of micro and macronutrients applied as liquid throughout the growth of the crop, may actually give the grower more control of his crop. To achieve optimum production, the grower can adjust nutrient levels to compensate for other environmental factors during the growing season. The absorption of mineral ions is dependent on a number of factors in addition to weather conditions. These include the cation exchange capacity or CEC and the pH or relative amount of hydrogen (H+) or hydroxyl ions (OH-) of the growing medium, and the total alkalinity of the irrigation water.

CEC or Cation Exchange Capacity

The Cation Exchange Capacity refers to the ability of the growing medium to hold exchangeable mineral elements within its structure. These cations include ammonium nitrogen, potassium, calcium, magnesium, iron, manganese, zinc and copper. Peat moss and mixes containing bark, sawdust and other organic materials all have some level of cation exchange capacity.

Meaning of pH

The term pH refers to the alkalinity or acidity of a growing media water solution. This solution consists of mineral elements dissolved in ionic form in water. The reaction of this solution whether it is acid, neutral or alkaline will have a marked effect on the availability of mineral elements to plant roots. When there is a greater amount of hydrogen H+ ions the solution will be acid (<7.0). If there is more hydroxyl OH- ions the solution will be alkaline (>7.0). A balance of hydrogen to hydroxyl ions yields a pH neutral soil (=7.0). The range for most crops is 5.5 to 6.2 or slightly acidic. This creates the greatest average level for availability for all essential plant nutrients. Extreme fluctuations of higher or lower pH can cause deficiency or toxicity of nutrients.

The Elements of Complete Plant Nutrition

The following is a brief guideline of the role of essential and beneficial mineral nutrients that are crucial for growth. Eliminate any one of these elements, and plants will display abnormalities of growth, deficiency symptoms, or may not reproduce normally.

Macronutrients

Nitrogen is a major component of proteins, hormones, chlorophyll, vitamins and enzymes essential for plant life. Nitrogen metabolism is a major factor in stem and leaf growth (vegetative growth). Too much can delay flowering and fruiting. Deficiencies can reduce yields, causes yellowing of the leaves and stunt growth.

Phosphorus is necessary for seed germination, photosynthesis, protein formation and almost all aspects of growth and metabolism in plants. It is essential for flower and fruit formation. Low pH (<4) results in phosphate being chemically locked up in organic soils. Deficiency symptoms are purple stems and leaves; maturity and growth are retarded. Yields of fruit and flowers are poor. Premature drop of fruits and flowers may often occur. Phosphorus must be applied close to the plant's roots in order for the plant to utilize it. Large applications of phosphorus without adequate levels of zinc can cause a zinc deficiency.

Potassium is necessary for formation of sugars, starches, carbohydrates, protein synthesis and cell division in roots and other parts of the plant. It helps to adjust water balance, improves stem rigidity and cold hardiness, enhances flavor and color on fruit and vegetable crops, increases the oil content of fruits and is important for leafy crops. Deficiencies result in low yields, mottled, spotted or curled leaves, scorched or burned look to leaves.

Sulfur is a structural component of amino acids, proteins, vitamins and enzymes and is essential to produce chlorophyll. It imparts flavor to many vegetables. Deficiencies show as light green leaves. Sulfur is readily lost by leaching from soils and should be applied with a nutrient formula. Some water supplies may contain Sulfur.

Magnesium is a critical structural component of the chlorophyll molecule and is necessary for functioning of plant enzymes to produce carbohydrates, sugars and fats. It is used for fruit and nut formation and essential for germination of seeds. Deficient plants appear chlorotic, show yellowing between veins of older leaves; leaves may droop. Magnesium is leached by watering and must be supplied when feeding. It can be applied as a foliar spray to correct deficiencies.

Calcium activates enzymes, is a structural component of cell walls, influences water movement in cells and is necessary for cell growth and division. Some plants must have calcium to take up nitrogen and other minerals. Calcium is easily leached. Calcium, once deposited in plant tissue, is immobile (non-trans locatable) so there must be a constant supply for growth. Deficiency causes stunting of new growth in stems, flowers and roots. Symptoms range from distorted new growth to black spots on leaves and fruit. Yellow leaf margins may also appear.

Micronutrients

Iron is necessary for many enzyme functions and as a catalyst for the synthesis of chlorophyll. It is essential for the young growing parts of plants. Deficiencies are pale leaf color of young leaves followed by yellowing of leaves and large veins. Iron is lost by leaching and is held in the lower portions of the soil structure. Under conditions of high pH (alkaline) iron is rendered unavailable to plants. When soils are alkaline, iron may be abundant but unavailable. Applications of an acid nutrient formula containing iron chelates, held in soluble form, should correct the problem.

Manganese is involved in enzyme activity for photosynthesis, respiration, and nitrogen metabolism. Deficiency in young leaves may show a network of green veins on a light green background similar to an iron deficiency. In the advanced stages the light green parts become white, and leaves are shed. Brownish, black, or grayish spots may appear next to the veins. In neutral or alkaline soils plants often show deficiency symptoms. In highly acid soils, manganese may be available to the extent that it results in toxicity.

Boron is necessary for cell wall formation, membrane integrity, and calcium uptake and may aid in the translocation of sugars. Boron affects at least 16 functions in plants. These functions include flowering, pollen germination, fruiting, cell division, water relationships and the movement of hormones. Boron must be available throughout the life of the plant. It is not trans located and is easily leached from soils. Deficiencies kill terminal buds leaving a rosette effect on the plant. Leaves are thick, curled and brittle. Fruits, tubers and roots are discolored, cracked and flecked with brown spots.

Zinc is a component of enzymes or a functional cofactor of a large number of enzymes including auxins (plant growth hormones). It is essential to carbohydrate metabolism; protein synthesis and internodal elongation (stem growth). Deficient plants have mottled leaves with irregular chlorotic areas. Zinc deficiency leads to iron deficiency causing similar symptoms. Deficiency occurs on eroded soils and is least available at a pH range of 5.5 - 7.0. Lowering the pH can render zinc more available to the point of toxicity.

Copper is concentrated in roots of plants and plays a part in nitrogen metabolism. It is a component of several enzymes and may be part of the enzyme systems that use carbohydrates and proteins. Deficiencies cause die back of the shoot tips, and terminal leaves develop brown spots. Copper is bound tightly in organic matter and may be deficient in highly organic soils. It is not readily lost from soil but may often be unavailable. Too much copper can cause toxicity.

Molybdenum is a structural component of the enzyme that reduces nitrates to ammonia. Without it, the synthesis of proteins is blocked and plant growth ceases. Root nodule (nitrogen fixing) bacteria also require it. Seeds may not form completely, and nitrogen deficiency may occur if plants are lacking molybdenum. Deficiency signs are pale green leaves with rolled or cupped margins.

Chlorine is involved in osmosis (movement of water or solutes in cells), the ionic balance necessary for plants to take up mineral elements and in photosynthesis. Deficiency symptoms include wilting, stubby roots, chlorosis (yellowing) and bronzing. Odors in some plants may be decreased. Chloride, the ionic form of chlorine used by plants, is usually found in soluble forms and is lost by leaching. Some plants may show signs of toxicity if levels are too high.

Nickel has just recently won the status as an essential trace element for plants according to the Agricultural Research Service Plant, Soil and Nutrition Laboratory in Ithaca, NY. It is required for the enzyme urease to break down urea to liberate the nitrogen into a usable form for plants. Nickel is required for iron absorption. Seeds need nickel in order to germinate. Plants grown without additional nickel will gradually reach a deficient level at about the time they mature and begin reproductive growth. If nickel is deficient plants may fail to produce viable seeds.

Sodium is involved in osmotic (water movement) and ionic balance in plants.

Cobalt is required for nitrogen fixation in legumes and in root nodules of non-legumes. The demand for cobalt is much higher for nitrogen fixation than for ammonium nutrition. Deficient levels could result in nitrogen deficiency symptoms.

Silicon is found as a component of cell walls. Plants with supplies of soluble silicon produce stronger, tougher cell walls making them a mechanical barrier to piercing and sucking insects. This significantly enhances plant heat and drought tolerance. Foliar sprays of silicon have also shown benefits reducing populations of aphids on field crops. Tests have also found that silicon can be deposited by the plants at the site of infection by fungus to combat the penetration of the cell walls by the attacking fungus. Improved leaf erectness, stem strength and prevention or depression of iron and manganese toxicity have all been noted as effects from silicon. Silicon has not been determined essential for all plants but may be beneficial for many.

Phosphate Solubilizing Bacteria

To meet the demand of overgrowing population it is the need of agrarian community to enhance the yield and future food supply. To overcome these problems, efforts needed to focus the soil biological system and the agro-ecosystem for better understanding the complex processes and their interactions for governing the stability of agricultural land.

The green revolution has been proved the most intellectual human activities contributing global food security and, consequently, changes the face of developing countries, such as India, from being food-deficient to having a food surplus. In present conditions there is an urgent need of second green revolution to increase the food production by around 50% in coming next 20 years to fulfill the demand of increasing population pressure.

Chemical fertilizers, such as water-soluble phosphatic (WSP) fertilizers have played a significant role in the green revolution to rectify the phosphorus deficiencies. However, excessive use of chemical pesticides arises soil health issues and beyond certain limit the yield plateau get declined. Thus it becomes clear that conventional agricultural practices cannot sustain the production base, for too long; while, to augment crop productivity agronomists have to rely on chemical fertilizers. In this context, after nitrogen, phosphorus is an essential plant nutrient whose deficiency marked the high yield. The phosphorous is present only in micro molar or lesser quantities in the earth's crust and is highly reactive with other elements in the soil.

Phosphorous is a one of the most abundant metallic elements found in the earth's crust and is present in the soils in both inorganic and organic forms. It is utilized or absorbed by the plants in inorganic form i.e. in orthophosphate (H_2PO_4- and HPO_4^{2-}). It has a key role in metabolic processes such as photosynthesis, energy transfer, signal transduction, nitrogen fixation in legumes, crop quality and resistance to plant diseases are the main features associated with phosphorous nutrition.

Phosphorus being a structural component of many coenzymes, phospho-proteins, phospholipids also forms a part of the genetic memory "DNA" of all living organisms. It involved in transfer and storage of energy which used for growth and reproduction. Phosphorus plays a lead role in especially in photosynthesis, carbon metabolism, and membrane formation also the vital role in elongation of root, proliferation, and phosphorous deficiency affects root architecture. A major portion of phosphorus absorbed by the plant is accumulated in grain in the form of phytic acid which becomes unavailable to plants and its deficiency negatively affects grains yield.

Tropical and subtropical regime has acidic soil considered as extremely deficient in phosphorus with high phosphorus sorption (fixation) capacities. On average, most mineral nutrients in soil solution are present in millimolar amounts but phosphorus is present only in micromolar or lesser quantities. The low levels of phosphorus are due to high reactivity of soluble phosphate with other elements.

A number of heterotrophic microorganisms excreting organic acids which solubilized P that chelate cationic partners of P ions and release the P directly into solution. These phosphate solubilizing bacteria (PSB) are being used as bio fertilizer since 1950s. Microbial inoculant assimilates soluble P, and prevents it from adsorption or fixation. These microorganisms influences soil fertility through various processes viz. decomposition, mineralization and release of nutrients. Microorganisms enhance the P availability to plants through solubilization of inorganic form of P to in available form. Hence, microbial inoculants are used as an alternate source, which are both economic as well as ecofriendly. A continued exploration of the natural biodiversity of soil microorganisms and the optimization of microbial interactions in the rhizosphere represents a prerequisite step to develop the more efficient microbial inoculants with phosphorus-solubilizing ability.

Phosphate –Plant Interaction

Phosphorus is one the major nutrient limiting plant growth. It has diverse role in plant nutrition and promotes the development of deeper roots. The soil that is rich in phosphorus constitutes about 0.05% (w/w) phosphorus but only one tenth of this is available to plants. Most of the P (95-99%) present in the soil in the insoluble form and hence cannot be utilized by the plants due to chemical fixation in the soil and it's interaction with other metallic elements that are present in the rhizospheric area. To increase the availability of phosphorus for plants, large amounts of fertilizer is used on a regular basis. But the continuous application of fertilizer P is rapidly transferred to the insoluble forms and thus there is a need of phosphate solubilizing microorganism to make the P in available form to the plants.

Phosphate Solubilizing Micro-Organisms

Naturally occurring rhizospheric phosphorus solubilizing microorganism (PSM) dates back to 1903. Number of microbial species plays key role in P solubilization these includes bacteria, fungi, actinomycetes and even algae. Bacteria are predominant amongst them and proved more effective in phosphorus solubilization than fungi. In addition to Pseudomonas and Bacillus, other bacteria reported as P solubilizes these are Rhodococcus, Arthrobacter, Serratia, Chryseobacterium, Phyllobacterium etc., Azotobacter, Xanthomonas, Enterobacter, Pantoea, and Klebsiella. Several halophilic bacteria Kushneria sinocarni have also been isolated from the sediment of Daqiao saltern on the eastern coast of China, which may be useful in stress conditions; salt affected agricultural soils. Among the whole microbial population in soil, PSB constitute 1 to 50 %, while phosphorus solubilizing fungi (PSF) are only 0.1 to 0.5 % in P solubilization potential which includes Penicillium and Aspergillus, Rhizoctonia solani, Trichoderma.

Moreover, fungi in soils are able to traverse long distances more easily than bacteria and hence, may be more important to P solubilization in soils. Generally, the P-solubilizing fungi produce more acids compared to bacteria and thus lead more P-solubilizing activity. Among the yeasts, Yarrowia lipolytica has potential to solubilize phosphate. Algae such as cyanobacteria and mycorrhiza have been also reported for P solubilization activity.

Mechanism of Phosphate Solubilization

The mechanism of P solubilization that is employed mostly by soil microorganisms includes: (1) release of complex compounds e.g. organic acid anions, siderophores, protons, hydroxyl ions, CO_2, (2) liberation of extracellular enzymes or it also referred as biochemical P mineralization and (3) the release of P during the degradation of substrate. Thus, microorganisms have key role in the soil P cycle i.e. precipitation, sorption–desorption, and mineralization.

Inorganic Phosphate Solubilization

Microorganism plays an important role in P solubilization through secretion of organic acid production either by: (i) lowering the pH, or (ii) through chelation reaction of cations bound to P (iii) by competing with P for adsorption sites on the soil.

The lowering in pH of the medium suggests the secretion of organic acids by the P solubilizing microorganisms via direct oxidation pathway that occurs on the outer face of the cytoplasmic membrane. When P is applied to soil it get interact with other metallic elements such as Fe, Al and Ca ions which makes the P unavailable to plants through the formation of ferrous phosphate, aluminium phosphate, calcium phosphate etc. and the release of organic acids by PSM leads the chelation reaction and because of this the bound P to other metallic elements get freed and becomes available to plants. The prominent acids that are released by PSM in the solubilization of insoluble P are gluconic acid, oxalic acid, citric acid, lactic acid, tartaric acid and aspartic acid etc.

The another mechanism is the production of H_2S, which react with ferric phosphate to yield ferrous sulphate with the release of phosphate. It could be because of the activity of PSM occurs as a consequence of microbial sulphur oxidation, nitrate production and CO_2 formation. These processes ultimately leads the formation of inorganic acids like sulphuric acid.

Organic Phosphate Solubilization

Phosphorus can be released in the soil from organic compounds by three groups of enzymes: (1) Nonspecific phosphatases, which leads dephosphorylation of phosphoester or phosphoanhydride bonds in organic matter, (2) Phytases, which mostly release P which is intact in the form of phytic acid, and (3) Phosphonatases and C–P Lyases, the phosphonates degrading enzyme enzymes that perform C–P cleavage in organophosphates. Availability of organic phosphate compounds for plant nutrition could be a limitation because as phosphorous is highly reactive it will interact with other metallic elements that are present in the soil in the rhizospheric area and becomes unavailable to plants which retard the plant growth and subsequently crop yield. Therefore, the capability of enzymes to perform the desired function in the rhizosphere is a crucial aspect for their effectiveness in plant nutrition.

Role of Microbial Exopolysaccharides in Phosphate Solubilization

Recently the role of polysaccharides in the microbial mediated solubilization of P was assessed by Yi et al. (2008). Microbial exopolysaccharides (EPSs) are polymers that mainly consist of carbohydrates excreted by some bacteria and fungi onto the outside of their cell walls. Bacterial strains such as Enterobacter sp. (EnHy-401), Arthrobacter sp. (ArHy-505), Azotobacter sp. (AzHy-510) and Enterobacter sp. (EnHy-402), has the ability to solubilize TCP (tri calcium phosphate).

Phosphate Solubilizing Bacteria as Plant Growth Promoters

There are several reports on plant growth promotion by bacteria that have the ability of solubilize inorganic and/or organic P from soil after their inoculation in soil or plant seeds. It was reported that a strain of Burkholderia cepacia, commercially used as bio fertilizer in Cuba which display significant mineral phosphate solubilization and moderate phosphatase activity, also improve the yield of tomato, potato, onion, banana, coffee etc. Inoculation with two strains of Rhizobium leguminosarum selected for their P solubilization ability has been shown to improve root colonization and growth promotion and to increase significantly the P concentration in lettuce and maize. Also a strain of Pseudomonas putida stimulates the growth of roots and shoots and increased $32P$-labeled phosphate uptake in canola. Co-inoculation of Pseudomonas striata and Bacillus polymyxa strains showing phosphate solubilizing activity, with a strain of Azospirillium brasilense, resulted in significant increase in grain and dry matter yields, with a concomitant increase in N and P uptake. Several studies have shown that PSB interacts with the vesicular arbuscular mycorrhizae (VAM) by releasing phosphate ions in the soil, which causes synergistic interaction that allows for better exploitation of poorly soluble P sources. Phylazonit-M is the commercial bio fertilizer which contains the mixtures of different bacterial cultures such as Bacillus megaterium, Azotobacter chroococcum increases N and P supply to the plants similarly, other product named "KYUSEI EM", a mixed inoculums of lactic acid bacteria, the organic acid lactic acid being the agent for the mineral phosphate solubilization.

Genetic Engineering of Phosphate Solubilizing Microorganisms

Some of the genes have been reported involved in mineral and organic phosphate solubilization has been so far isolated and characterized. Genetic manipulation of these genes followed by their expression in selected rhizobacterial strains leads a promising perspective for obtaining PSM strains with higher phosphate solubilizing capacity, and thus it become more effective as agricultural inoculants. The initial achievement was achieved by Goldstein and Liu from the Gram negative bacteria Erwinia herbicola through cloning of P solubilization genes. Similarly, the nap A phosphatase gene from the soil bacterium Morganella morganii was transferred to Burkholderia cepacia IS-16, a strain used as a bio fertilizer, using the broad host range vector pRK293. Introduction of P solubilization genes in natural rhizosphere bacteria is a candidate approach for the improvement of microbial capacity.

Fourteen different nonspecific acid phosphatase encoding genes have been isolated from different bacterial species using different expression cloning system. Sequence analysis of the cloned phosphatase genes allowed the classification into three different families: class A, class B, and class C phos- phatases. Several other phosphatase genes have been isolated from Escherichia coli. These include: ushA, which encodes a 5'-nucleotidase; agp, which encodes an acid glucose-1- phosphatase and cpdB, encoding the 2'-3' cyclic phosphodiesterase. Despite the difficulties, significant progress has been made for obtaining genetically engineered microorganisms for the agricultural use.

In conclusion, phosphorus is an essential element in crop nutrition. Continuous application of chemical based P fertilizers compelled us to find a sustainable approach for efficient P availability in agriculture to meet the over growing demand of food. Soil microorganisms plays varied role that affect the transformation of P and thus influence the availability of P to plant roots.

Table: Biodiversity of P Solubilizing Microorganisms.

Groups	Examples
N$_2$ fixing biofertilizers	
Free living	Azotobacter, Beijerinkia, Clostridium, Klebsiella, Anabaena and Nostoc
Symbiotic	Rhisobium, Frankia and Anabaena azollae
Associative symbiotic	Azospirillum
P solubilizing bifrertilizers	
Bacteria	Bacillus megaterium var. phosphaticum, Bacillus circulans and Pseudomonas striata
Fungi	Penicillium sp. and Aspergillus awamori
P mobilizing biofertilizers	
Arboscular mycorhiza	Glomus sp., Gigaspora sp., Acaulospora sp., and Sclerocystis sp.
Ectomycorrhiza	Laccaria sp., Boletus sp., and Amanita sp.
Ericoidmycorrhiza	Pezizella
Orchid mycorrhiza	Rhizoctonia solani
Biofertilizers for micro nutrients	
Silicate and Zinc solubilizers	Bacillus sp.
Plant growth promoting rhizobacteria	
Pseudomonas	Pseudomonas fluorescens

The use of efficient PSM (phosphatesolubilizing microorganisms), opens up a new horizon for better crop productivity and for greater yield performance without affecting the soil health. Phosphate solubilizing bacteria play an important role in the plant nutrition through increase in P uptake by the plants and their use as PGPR is an important contribution to bio fertilization of agricultural crops. Therefore, steps should be taken for extensive and consistent research for the identification and characterization of PSM with greater efficiency for their ultimate application under field conditions. It becomes the responsibility of soil scientists and microbiologists how soil P could be improved without applying the chemical based phosphatic fertilizers under different agroclimatic regions.

Genetic engineering of the phosphate solubilizing character must eventually be directed to the chromosomal integration of the gene for higher stability of the character and to avoid horizontal transfer of the inserted gene in soil. This strategy would also prevent the risk of metabolic load caused by the presence of the plasmid in the bacterial cell.

Thus the exploitation of the candidate soil microorganisms which play an important role in the mobilization of soil P and understanding the mechanism of phosphate solubilization, phosphate plant interaction and their contribution to the cycling of P in soil plant systems is essential for the development of sustainable agriculture to forward and accomplished our movement from a green revolution to an evergreen revolution.

References

- Chen, Hongwei; An, Jing; Wei, Shuhe; Gu, Jian; Liang, Wenju (2015). "Spatial Patterns and Risk Assessment of Heavy Metals in Soils in a Resource-Exhausted City, Northeast China". PLOS ONE. 10 (9): e0137694. doi:10.1371/journal.pone.0137694

- Edaphology-1800: winefrog.com, Retrieved 12 March 2018

- Ziadat, Feras Mousa; Yeganantham, Dhanesh; Shoemate, David; Srinivasan, Raghavan; Narasimhan, Balaji; Tech, Jaclyn (2015). "Soil-Landscape Estimation and Evaluation Program (SLEEP) to predict spatial distribution of soil attributes for environmental modeling". International Journal of Architectural and Biological Engineering. 8 (3): 158–172. doi:10.3965/j.ijabe.20150803.1270

- Agricultural-soil, agricultural-and-biological-sciences: sciencedirect.com, Retrieved 10 July 2018

Chapter 5

Soil Properties

The properties of the soil such as soil texture, structure, resistivity, bulk density, etc. play a crucial role in ecosystem services. The study of the soil requires an understanding of its varied properties, such as soil structure, soil resistivity, soil texture, pore space of soil, thermal properties, etc. which have been carefully elaborated in the chapter.

Soil Structure

Soil structure is the result of the spatial arrangement of the solid soil particles and their associated pore space. Aggregation mainly depends on the soil composition and texture, but is also strongly influenced by other factors such as biological activity, climate, geomorphic processes or the action of fire. Structure is a typical morphological soil property, which allows differentiating soil of geological material. Because of its importance, structure is a property commonly described in soil studies.

Organic and mineral soil particles are not isolated from each other, but form structural aggregates (also called "peds"). In 1961, Blackmore and Miller observed how the Ca-montmorillonite may be arranged in groups of four or five particles, depending on various soil characteristics.

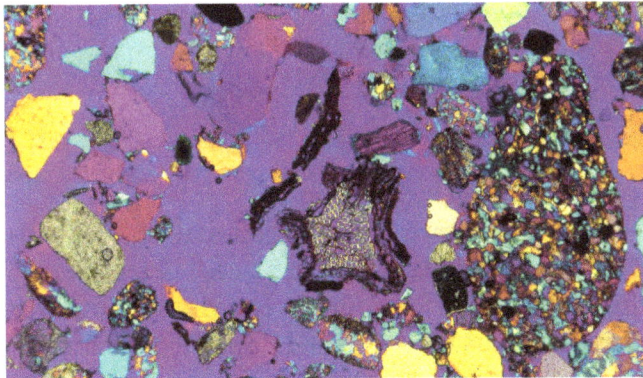

Thin section of a surface sandy soil under cross polarized light showing sand grains and cellular plant material.

The fact that soil particles do not form a continuous and compact mass, but are associated, involves an interconnected pore space, makes possible the development of life in the soil. The volume formed by pores, channels, chambers and cracks allows the movement of fluids (air and water) in the soil, providing a favorable environment for microbial activity and facilitating root growth of plants.

Some authors consider that more than one property, structure is a state of soil, because when dry, it becomes clear, but if it is wet, the soil becomes massive, no cracks are distinguishable, and structure disappears.

Texture, biological activity and a number of physicochemical conditions allow the aggregation of soil particles. The predominance of one or other process creates various types of structure. Aggregation is strongly conditioned by colloids (clay and organic matter) and soil cementing substances (carbonates, sesquioxides, etc.), which coat solid particles, including them in groups (aggregates). If the proportion of colloids or cementing substances is too low, solid particles remain dispersed. Flocculation of colloids gives rise to the co-precipitation of colloidal particles (clay and organic matter), forming microaggregates (<250 µm), which then evolve resulting in macroaggregates (>250 µm). In the formation of small fabric units (cluster and domains), the inorganic bonds are the most important, while in the aggregate stabilization the organic ones play a more relevant role (humic cements).

Flocculation induced by cations in the soil solution plays an important role in the development of aggregates. Calcium and magnesium (in calcareous soils) or iron and aluminium (in acid soils) favour the formation of stable aggregates. In contrast, monovalent cations as sodium contribute to dispersion of aggregates. Also, cementing agents as calcium carbonate (in calcareous soils) or iron oxides (acid soils) may enhance soil aggregation. In the formation of macroaggregates, biological agents are also involved, as plants (roots), animals (earthworms, arthropods, etc.), microorganisms (bacteria and, especially, fungi) are also important.

Fragment of calcium carbonate from an unearthed petro calcic horizon.

The degree of development of the structure and aggregate stability depends on the type of particles present and the forces of attraction/repulsion taking place. This can lead to particle packing or aggregate formation. Packing is important when the forces of attraction / repulsion are negligible in the absence of electric charge (such us between sand particles). In sandy soils, the surface tension of the film of water adsorbed on the surface of the grains may cause a certain binding capacity.

Aggregate Stability

Soil structure is not a stable parameter; it may vary depending on weather conditions, management, soil processes, etc. In general, the most important causes of the degradation of soil structure are:

- Expansion of swelling clays (montmorillonite type) during wet periods.

- Rain, especially if it results in a violent dilution of cations, which promotes flocculation of the colloids.

- Loss of organic matter (common in cropped or eroded soils).

- Acidification, resulting in destabilization of microaggregates.

Types of Soil Structure:

There are four principal forms of soil structure:

(a) Plate-like:

In this structural type of aggregates are arranged in relatively thin horizontal plates. The horizontal dimensions are much more developed than the vertical. When the units are thick, they are called platy, and when thin, laminar (Fig.).

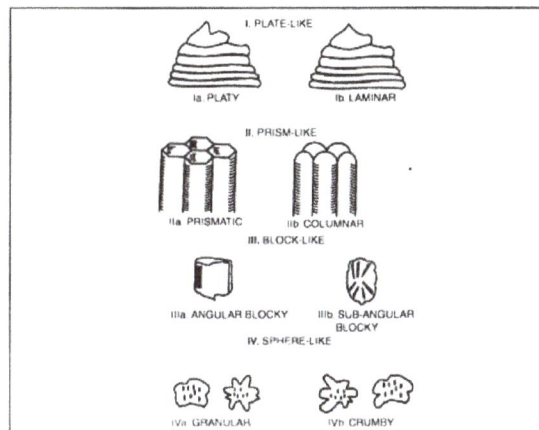

Fig. Type of soil structure.

Platy structure is most noticeable in the surface layers of virgin soils but may be present in the subsoil. Although most structural features are usually a product of soil forming forces, the platy type is often inherited from the parent material, especially those laid down by water.

(b) Prism-like:

The vertical axis is more developed than horizontal, giving a pillar-like shape. When the top of such a ped is rounded, the structure is termed as columnar, and when flat, prismatic. They commonly occur in sub-soil horizons in arid and semi-arid regions.

(c) Block-like:

All these dimensions are about the same size and the peds are cube-like with flat or rounded faces. When the faces are flat and the edges sharp angular, the structure is named as angular blocky. When the faces and edges are mainly rounded it is called sub angular blocky. These types usually are confined to the sub-soil and characteristics have much to do with soil drainage, aeration and root penetration.

(d) Spheroidal (Sphere-like):

All rounded aggregates (peds) may be placed in this category, although the term more properly refers to those not over 0.5 inch in diameter. Those rounded complexes usually lie loosely and separately.

When wetted, the intervening spaces generally are not closed so readily by swelling as may be the case with a blocky structural condition. Therefore in sphere-like structure infiltration, percolation and aeration are not affected by wetting of soil. The aggregates of this group are usually termed as granular which are relatively less porous; when the granules are very porous, the term used is crumby.

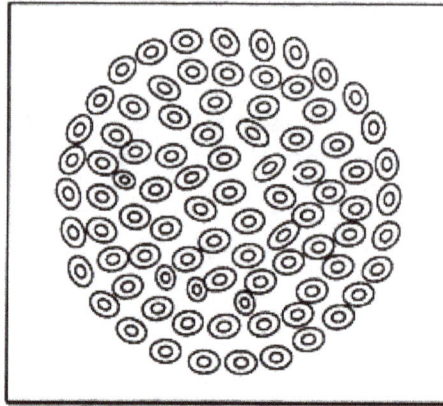

Fig. (a)Example of Sphere like soil structure.

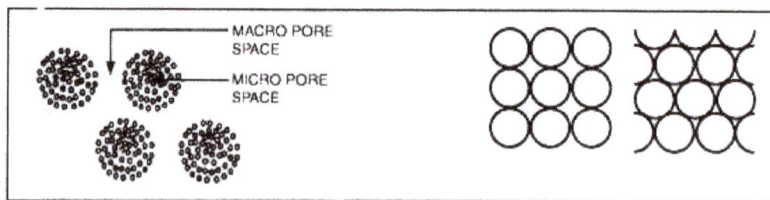

Fig. (b) Arrangement of particles in Sphere-like soil structure.
Fig. (c) Arrangement of Spherical particles in two ways

Classes of Soil Structure

Each primary structural type of soil is differentiated into 5 size-classes depending upon the size of the individual peds.

The terms commonly used for the size classes are:

1. Very fine or very thin

2. Fine or thin

3. Medium

4. Coarse or thick

5. Very coarse or very thick.

The terms thin and thick are used for platy types, while the terms fine and coarse are used for other structural types.

Grades of Soil Structure

Grades indicate the degree of distinctness of the individual peds. It is determined by the stability of the aggregates. Grade of structure is influenced by the moisture content of the soil. Grade also depends on organic matter, texture etc.

Four terms commonly used to describe the grade of soil structure are:

1. Structure-less:

There are no noticeable peds, such as conditions exhibited by loose sand or a cement-like condition of some clay soils.

2. Weak Structure:

Indistinct formation of peds which are not durable.

3. Moderate Structure:

Moderately well-developed peds which are fairly distinct.

4. Strong Structure:

Very well-formed peds which are quite durable and distinct. For naming a soil structure the sequence followed is grade, class and type; for example, strong coarse angular blocky (soil structure).

Examples of Sphere-like Soil Structure

Grade	Class	Type	Structural Name
1. Strong	fine	granular	strong fine granular
2. Moderate	very coarse	granular	moderate very coarse granular

Often compound structures are met within the soil under natural conditions. For example, large prismatic types may break into medium blocky structure, constitute the compound structure.

Formation of Soil Structure

The mechanism of structure (aggregate) formation is quite complex. In aggregate formation a number of primary particles such as sand, silt and clay are brought together by the cementing or binding effect of soil colloidal clay, iron and aluminium hydroxides and organic matter.

The mineral Colloids (colloidal clay) by virtue of their properties of adhesion and cohesion, stick together to form aggregates. Sand and silt particles cannot form aggregates as they do not possess the power of adhesion and cohesion.

The amount and nature of colloidal clay influence the formation of aggregates. The greater the amount of clay in a soil, the greater is the tendency to form aggregates. Clay particles smaller than 0.001 mm aggregate very readily. So also clay minerals that have high base exchange capacity form aggregate more readily than those which have a low base exchange capacity. Iron and aluminium hydroxides act as cementing agent is binding the soil particles together. These are also responsible for forming aggregates by cementing sand and silt particles.

Organic matter plays an important part in forming soil aggregates. During decomposition of organic matter, humic acid and other sticky materials are produced which helps to form aggregate. Some fungi and bacteria taking part in the decomposition have also been found to have a cementing effect.

Another view of structure formation is that clay particles adsorbed by humus forming a clay-humus complex. It seems that humus absorbs both cations and anions. In normal soil, calcium is the predominant cation and forms calcium humate in combination with humus.

Factors Affecting Soil Structure:

The development of structure in arable soil depends on the following factors:

1. Climate:

Climate has considerable influence on the degree of aggregation as well as 011 the type of structure. In arid region, there is very little aggregation of primary particles. In semi- arid regions, the degree of aggregation is greater than arid regions.

2. Organic Matter:

Organic matter improves the structure of a sandy soil as well as of a clay soil. In a case of sandy soil, the sticky and slimy material produced by the decomposing organic matter and the associated microorganism cement the sand particles to form aggregates. In the case of clayey soil, it modifies the properties of clay by reducing its cohesive power. This helps making clay more crumby.

3. Tillage:

Cultivation implements break down of large clods into smaller fragments and aggregates. For obtaining good granular and crumby structure, optimum moisture content in the soil is necessary. If the moisture content is too high it will form large clods on drying. If it is too low, some of the existing aggregates will be broken down.

4. Plant Roots:

Large numbers of granules remain attached to roots and root hairs which help to develop crumb structure. Plant root secretions may also act as cementing agents in binding the soil particles. The plant roots, on decay, may also bring about granulation due to the production of sticky substances.

5. Soil Organism:

Among the soil fauna, small animals like earthworms, moles and insects etc., that burrow in the soil are the chief agents that take part in the aggregation of finer particles.

6. Fertilizers:

Fertilizer like Sodium nitrate destroys granulation by reducing the stability of aggregates. Few fertilizers, for example, Calcium Ammonium nitrate, help in development of good structures.

7. Wetting and Drying:

Wren a dry soil is wetted, the soil colloids swell on absorbing water. On drying, shrinkage produced strains in the soil mass give rise to cracks which break it up into clods and granules of various sizes.

Effects of Soil Structure on Other Physical Properties of Soil

Soil structure brings change in other physical properties of soil like porosity, temperature, density, consistency and colour.

1. Porosity:

Porosity of a soil is easily changed. In plate-like structure pore spaces are less whereas in crumby structure pore spaces are more.

2. Temperature:

Crumby structure provides good aeration and percolation in the soil. Thus, these characteristics help in keeping optimum temperature in comparison to plate-like structure.

3. Density:

Bulk density varies with the total pore space present in the soil. Structure chiefly influences pore spaces. Platy structure with less total pore spaces has high bulk density whereas crumby structure with more total pore spaces has low bulk density.

4. Consistence:

Consistence of soil also depends on structure. Plate-like structure exhibits strong plasticity.

5. Colour:

Bluish and greenish colours of soil are generally due to poor drainage of soil. Platy structure normally hinders free drainage.

Structural Management of Soils

(a) Coarse-textured Soil:

Sandy soils are commonly too loose and lack the capacity to adsorb and hold sufficient moisture and nutrients. They lack fertility and water- holding capacity. There is only one practical method of improving the structure of such soil- the addition of organic matter. Organic matter will not only act as a binding agent for the particles but will also increase the water-holding capacity. Sod-crops, for example, corn, blue grass etc., also help in improving the structural condition of sandy soils.

(b) Fine-textured Soil:

The structural management of a clay soil is difficult than sandy soil. In clay, plasticity and cohesion are high because of the presence of large amount of colloidal clay. When such a soil is tilled when wet, its pore space becomes much reduced, it becomes practically impervious to air and water and it is said to be puddled. When a soil in this condition dries, it usually becomes hard and dense.

The tillage of clay soil should be done at right moisture stage. If ploughed too wet, the structural aggregates are broken down and an un-favorable structure results. On the other hand, if ploughed too dry, big clods are turned up which are difficult to work. The granulation of fine-textured soil should be encouraged by the incorporation of organic matter. Growing of sod-crops also improves granulation in the soil.

(c) Rice Soil:

Puddling of the soil is generally beneficial to the production of rice. In preparation for the planting of rice, the soil is flooded with water and then puddled by intensive tillage. Puddling destroys the structural aggregates. Rice seedling is transplanted into the freshly prepared mud.

Such soil management helps control weeds and also reduces the rate of water movement down (percolation) through the soil. This is important to maintain standing water in the rice throughout the growing season. By reducing water percolation, puddled soil markedly decreases the amount of water needed to produce a rice crop.

Semi-aquatic characteristics of the rice plant account for its positive response to a type of soil management that destroys aggregate. Rice survives flooded conditions because oxygen moves downward inside the stem of the plant to supply the roots. This characteristic permits rice to stand well in the water-logging condition. Rice can be grown successfully on un-puddled but flooded soil.

Soil Resistivity

The measure of the resistance offered by the soil in the flow of electricity is called the soil resistivity. The resistivity of the soil depends on the various factors likes' soil composition, moisture, temperature, etc. Generally, the soil is not homogenous, and their resistivity varies with the depth. The soil having a low resistivity is good for designing the grounding system. The resistivity of the soil is measured in ohmmeter or ohm-centimeters.

The resistivity of the soil mainly depends on its temperature. When the temperature of the soil is more than 0°, then its effect on soil resistivity is negligible. At 0° the water starts freezing and resistivity increases. The magnitude of the current also affects the resistivity of the soil. If the magnitude of current dissipated in the soil is high, it may cause significant drying of soil and increase its resistivity.

The resistivity of the soil varies with the depth. The lower layers of the soil have greater moisture content and lower resistivity. If the lower layer contains hard and rocky layers, then their resistivity may increase with the depth.

Measurement of Soil Resistivity

The resistivity of the soil is usually measured by the four spike methods. In this method the four spikes arranged in the straight line are driven into the soil at equal distance. A known current is passed between electrode C1 and C2 and potential drop V is measured across P1 and P2. The current I developed an electric field which is proportional to current density and soil resistivity. The voltage V is proportional to this field.

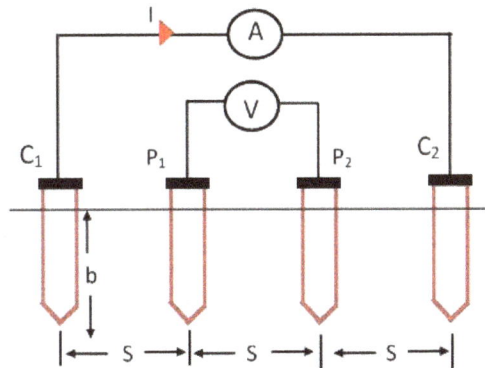

Measurement of Soil Resistivity

The soil resistivity is proportional to the ratio of the voltage V and current I and is given as

$$\rho = \frac{\dfrac{4\pi SV}{I}}{1 + \dfrac{2S}{\sqrt{S^2 + 4b^2}} - \dfrac{2S}{\sqrt{4S^2 + 4b^2}}}$$

Where ρ is the resistivity of the soil and their unit is ohmmeters. S is the horizontal space between the spikes in m and b is the depth of burial in metre.

If the measurement is to be carried out using the main supply, an isolating transformer should be connected between the main supply and the test setup. So that the result may not be affected by it.

Soil Texture

The inorganic material in soil is called mineral matter. Mineral matter began as rock that was weathered into small particles. Most soils have different sizes of mineral particles. These particles are called sand, silt, or clay, based on their size.

Sand is the largest of the mineral particles. Sand particles create large pore spaces that improve aeration. Water flows through the large pore spaces quickly. Soils with a high percentage of sand are generally well-drained. Sandy soils lack the ability to hold nutrients and are not fertile. Sandy soils also feel gritty to the touch.

Silt is a mid-size soil particle. It has good water-holding ability and good fertility characteristics. Silt feels like flour when dry and smooth like velvet when moist.

Clay is the smallest size soil particle. Clay has the ability to hold both nutrients and water that can be used by plants. It creates very small pore spaces, resulting in poor aeration and poor water drainage. Clay forms hard clumps when dry and is sticky when wet.

TABLE: Characteristics of Sand, Silt, and Clay

Characteristics	Sand	Silt	Clay
Looseness	Good	Fair	Poor
Air Space	Good	Fair to Good	Poor
Drainage	Good	Fair to Good	Poor
Tendency to Form Clods	Poor	Fair	Good
Ease of Working	Good	Fair to Good	Poor
Moisture-Holding Ability	Poor	Fair to Good	Good
Fertility	Poor	Fair to Good	Fair to Good

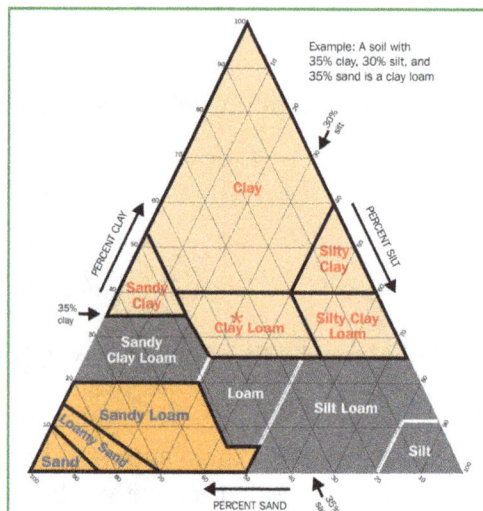

Figure. The textural triangle can be used to differentiate the several classes of soil.

Textural Classes

Soil texture is the proportion of the three sizes of soil particles that are present and the fineness or coarseness of a soil. Soil texture may be determined in one of two ways. The percentages of sand, silt, and clay may be tested in the lab. Once tested, the textural class of the soil can be determined by referring to the textural triangle (a device used to differentiate the several classes of soil). Soils with different amounts of sand, silt, and clay are given different names. For instance, a soil containing 40 percent sand, 20 percent clay, and 40 percent silt is called loam soil. The relative amounts of sand, silt, and clay may also be determined in the field using the ribbon method. Five textural classes may be determined using this method.

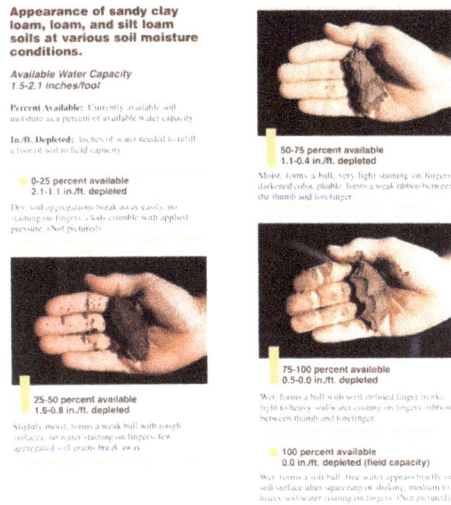

Appearance of sandy clay
loam, loam, and silt loam
soils at various soil moisture
conditions.

Available Water Capacity
1.5-2.1 inches/foot

Percent Available: Currently available soil
moisture as a percent of available water capacity.

In./ft. Depleted: Inches of water needed to refill
a foot of soil to field capacity.

0-25 percent available
2.1-1.1 in./ft. depleted

Dry, soil aggregations break away easily, no
staining on fingers, clods crumble with applied
pressure. (Not pictured)

50-75 percent available
1.1-0.4 in./ft. depleted

Moist, forms a ball, very light staining on fingers,
darkened color, pliable, forms a weak ribbon between
the thumb and forefinger.

25-50 percent available
1.5-0.8 in./ft. depleted

Slightly moist, forms a weak ball with rough
surfaces, no water staining on fingers, few
aggregated soil grains break away.

75-100 percent available
0.5-0.0 in./ft. depleted

Wet, forms a ball with well defined finger marks,
light to heavy soil/water coating on fingers, ribbon
between thumb and forefinger.

100 percent available
0.0 in./ft. depleted (field capacity)

Wet, forms a soft ball, free water appears briefly on
soil surface after squeezing or shaking, medium to
heavy soil/water coating on fingers. (Not pictured)

Figure. The ribbon method is used to determine five major textural classes of soil.

- Fine-textured—A ribbon of soil forms easily and remains long and flexible.

- Moderately fine-textured—A ribbon forms but breaks into pieces 3/4 to 1 inch long.

- Medium-textured—No ribbon forms. The sample breaks into pieces less than 3/4 inch long. The soil feels smooth and talc-like.

- Moderately coarse-textured—No ribbon forms. The sample feels gritty and lacks smoothness.

- Coarse-textured—No ribbon forms. The sample is composed almost entirely of gritty material and leaves little or no stain.

Table: Soil Textural Classes

Sand	Dry—Loose and single grained; feels gritty.
	Moist—Will form very easily—crumbled ball
	Sand: 85-100%, Silt: 0-15%, Clay: 0-10%
Loamy Sand	Dry—Silt and clay may mask sand; feels loose, gritty.
	Moist—Feels gritty; forms easily—crumbled ball; stains fingers slightly.
	Sand: 70-90%, Silt: 0-30%, Clay: 0-15%
Sandy Loam	Dry—Clods easily broken; sand can be seen and felt.
	Moist—Moderately gritty; forms ball that can stand careful handling; definitely stains fingers.
	Sand: 43-85%, Silt: 0-50%, Clay: 0-20%
Loam	Dry—Clods moderately difficult to break; somewhat gritty.
	Moist—Neither very gritty nor very smooth; forms a ball; stains fingers.
	Sand: 23-52%, Silt: 28-50%, Clay: 7-27%

Silt Loam	Dry—Clods difficult to break; when pulverized feels smooth, soft, and floury, shows fingerprints.
	Moist—Has smooth or slick buttery feel; stains fingers.
	Sand: 0-50%, Silt: 50-88%, Clay: 0-27%
Clay Loam	Dry—Clods very difficult to break with fingers.
	Moist—Has slight gritty feel; stains fingers; ribbons fairly well.
	Sand: 20-45%, Silt: 15-53%, Clay: 27-40%
Silty Clay Loam	Same as above, but very smooth.
	Sand: 0-20%, Silt: 40-73%, Clay: 27-40%
Sandy Clay Loam	Same as for Clay Loam.
	Sand: 45-80%, Silt: 0-28%, Clay: 20-35%
Clay	Dry—Clods cannot be broken with fingers without extreme pressure.
	Moist—Quite plastic and usually sticky when wet; stains fingers. (A silty clay feels smooth, a sandy
	clay feels gritty.)
	Sand: 0-45%, Silt: 0-40%, Clay: 40-100%

Soil Characteristics Related to Texture

The texture of a soil is important because it determines soil characteristics that affect plant growth. A few of these characteristics are water-holding capacity, permeability, and soil workability. Water-holding capacity is the ability of a soil to retain water. Most plants require a steady supply of water, and it is obtained from the soil. While plants need water, they also need air in the root zone. The ease with which air and water may pass through the soil is called permeability. Soil workability is the ease with which soil may be tilled and the timing of the work.

Soils with a larger percentage of sand are easier to work than soils with a larger percentage of clay. Clay soil tends to be tighter, making it more difficult to break up or cultivate, whereas sandy soil is looser. It also takes longer for a clay soil to dry after a rain than a sandy soil.

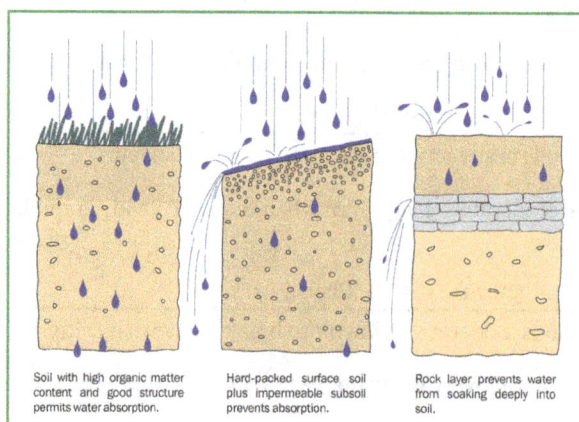

Soil with high organic matter content and good structure permits water absorption.

Hard-packed surface soil plus impermeable subsoil prevents absorption.

Rock layer prevents water from soaking deeply into soil.

Figure. Soil with good permeability allows water to be easily absorbed.

Because of the better drainage, a sandy soil can be worked sooner. In the case of wet clay soil, the producer or gardener must wait longer for the soil to dry sufficiently. Soil texture may limit which crops can be grown. For example, root crops, such as carrots and onions, perform best in a sandy soil because it is loose and allows the plant to expand. However, growth of some plants is stunted when growing in sandy soils because they lack the water- and nutrient-holding ability.

Pore Space of Soil

The pore space in a soil represents that part of the soil volume which has been occupied by air or water.

Pore space in a soil may be of the following three types:

(i) Macro pore space occurs between peds. They are more than 75 micron in diameter. Water and air rapidly pass through them. They are necessary for rapid drainage of the soil.

(ii) Mesopores are 75 to 30 microns in diameter. Relatively larger mesopores occur between the peds while relatively smaller ones occur within the peds. Mesopores are required for storage of water for crop growth.

(iii) Micro-pores are less than 30 microns in diameter. Water is held within the micro-pores tightly enough to be unavailable for plant growth.

Factors Affecting Pore Space Per cent of Soils

(i) Texture:

Total and Micro-pore spaces of soils increase when the clay percentage increases i.e. their texture becomes finer. The microspores ices of clayey soils are much more than those of the sandy soils, because clay particles unite to form soil aggregate within which micro-pore space occur.

(ii) Organic Matter:

Organic matter is decomposed by the soil microorganisms to form humus, which usually binds the soil particles to form aggregates. Besides this, humus also stimulates the growth of soil microorganisms, which mechanically bind the soil particles to form aggregates within which micro-pore spaces occur. Hence the micro and total pore spaces increase.

(iii) Nature of Crops and Cultivation:

Excessive cultivation of soils increases the oxidation of humus. Consequently, total and micro-pore spaces of soils are decreased if soils are more and more cultivated. Less cultivation is required for growing grasses, the roots of which bind the soil particles to form soil aggregates.

Crops like potato, maize etc. require the maximum amount of soil cultivation for soil aeration which increases the oxidation of humus. Hence grasses increase the percentage of pore space of the soil, whereas arable crops like maize, potato etc. decreases it.

(iv) Soil Depth:

Since plant roots and organic matter occur more in surface soil then in sub-soil, so pore space per cent of the surface soil is usually much more than that of the sub-soil.

Thermal Properties of Soil

The thermal properties of soils are a component of soil physics that has found importants uses in engineering, climatology and agriculture. These properties influence how energy is partitioned in the soil profile. While related to soil temperature, it is more accurately associated with the transfer of heat throughout the soil, by radiation, conduction and convection.

Main soil thermal properties:

Volumetric heat capacity, SI units: $Jm^{-3}K^{-1}$

Thermal conductivity, SI units: $W.m^{-1}K^{-1}$

Thermal diffusivity, SI units: $m^2.s$

Soil Temperature – Soil Air – Gaseous Exchange

Soil Temperature

Soil temperature is an important plant growth factor like air, water and nutrients. Soil temperature affects plant growth directly and also indirectly by influencing moisture, aeration, structure, microbial and enzyme activities, rate of organic matter decomposition, nutrient availability and other soil chemical reactions. Specific crops are adapted to specific soil temperatures. Apple grows well when the soil temperature is about 18°C, maize 25°C, potato 16 to 21°C, and so on.

Sources of Soil Heat

The sources of heat for soil are solar radiation, heat released during microbial decomposition of organic matter and respiration by soil organisms including plants and the internal source of heat is the interior of the Earth - which is negligible. The rate of solar radiation reaching the earth's atmosphere is called as solar constant and has a value of 2 cal cm^{-2} min^{-1}. Major part of this energy is absorbed in the atmosphere, absorbed by plants and also scattered. Only a small part of it reaches soil. Thermal energy is transmitted in the form of thermal infrared radiation from the sun across the space and through the atmosphere.

Factors Affecting Soil Temperature

The average annual soil temperature is about 1°C higher than mean annual air temperature. Soil temperature is influenced by climatic conditions. The factors that affect the transfer of heat through the atmosphere from sun affect the soil temperature also.

Environmental Factors

Solar radiation: The amount of heat received from sun on Earth's surface is 2 cal cm-2 min-1. But the amount of heat transmitted into soil is much lower. The heat transmission into soil depends on the angle on incident radiation, latitude, season, time of the day, steepness and direction of slope and altitude. The insulation by air, water vapour, clouds, dust, smog, snow, plant cover, mulch etc., reduces the amount of heat transferred into soil.

Soil Factors

a) Thermal (Heat) capacity of soil: The amount of energy required to raise the temperature by 1°C is called heat capacity. When it is expressed per unit mass, then it is called as specific heat. The specific heat of water is 1.00 cal g-1 where the specific heat of a dry soil is 0.2 cal g-1. Increasing water content in soil increases the specific heat of the soil and hence a dry soil heats up quickly than a moist soil.

b) Heat of vaporization: The evaporation of water from soil requires a large amount of energy, 540 kilocalories kg-1 soil. Soil water utilizes the energy from solar radiation to evaporate and thereby rendering it unavailable for heating up of soil. Also the thermal energy from soil is utilized for the evaporation of water, thereby reducing the soil temperature. This is the reason that surface soil temperatures will be sometimes 1 to 6°C lower than the sub-surface soil temperature. That is why the specific heat of a wet soil is higher than dry soil.

c) Thermal conductivity and diffusivity: This refers to the movement of heat in soils. In soil, heat is transmitted through conduction. Heat passes from soil to water about 150 times faster than soil to air. So the movement of heat will be more in wet soil than in dry soil where the pores will be occupied with air. Thermal conductivity of soil forming materials is 0.005 thermal conductivity units, and that of air is 0.00005 units, water 0.001 units. A dry and loosely packed soil will conduct heat slower than a compact soil and wet soil.

d) Biological activity: Respiration by soil animals, microbes and plant roots evolve heat. More the biological activity more will be the soil temperature.

e) Radiation from soil: Radiation from high temperature bodies is in short waves (0.3 to 2.2 µ) and that from low temperature bodies is in long waves (6.8 to 100 µ) Longer wavelengths have little ability to penetrate water vapour, air and glass and hence soil remains warm during night hours, cloudy days and in glass houses.

f) Soil colour: Colour is produced due to reflection of radiation of specific wavelengths. Dark coloured soils radiate less heat than bright coloured soils. The ratio between the incoming and outgoing (reflected energy) radiation is called albedo. The larger the albido, the cooler is the soil. Rough surfaced soil absorbs more solar radiation than smooth surface soils.

$$\text{Albido} = \frac{\text{Reflected energy}}{\text{Incidentergy}}$$

g) Soil structure, texture and moisture: Compact soils have higher thermal conductivity than loose soils. Natural structures have high conductivity than disturbed soil structures. Mineral soils have higher conductivity than organic soils. Moist soil will have uniform temperature over depth because of its good conductivity than dry soils.

h) Soluble salts: Indirectly affects soil temperature by influencing the biological activities, evaporation etc.

Soil Air

Soil air is a continuation of the atmospheric air. Unlike the other components, it is constant state of motion from the soil pores into the atmosphere and from the atmosphere into the pore space. This constant movement or circulation of air in the soil mass resulting in the renewal of its component gases is known as soil aeration.

Composition of Soil Air: The soil air contains a number of gases of which nitrogen, oxygen, carbon dioxide and water vapour are the most important. Soil air constantly moves from the soil pores into the atmosphere and from the atmosphere into the pore space. Soil air and atmospheric air differ in the compositions. Soil air contains a much greater proportion of carbon dioxide and a lesser amount of oxygen than atmospheric air. At the same time, soil air contains a far great amount of water vapour than atmospheric air. The amount of nitrogen in soil air is almost the same as in the atmosphere.

Composition of Soil and Atmospheric Air

	Percentage by volume		
	Nitrogen	Oxygen	Carbon Dioxide
Soil air	79.2	20.60	0.30
Atmospheric air	79.9	20.97	0.03

Factors Affecting the Composition of Soil Air:

1. Nature and condition of soil: The quantity of oxygen in soil air is less than that in atmospheric air.

The amount of oxygen also depends upon the soil depth.

The oxygen content of the air in lower layer is usually less than that of the surface soil. This is possibly due to more readily diffusion of the oxygen from the atmosphere into the surface soil than in the subsoil.

Light texture soil or sandy soil contains much higher percentage than heavy soil. The concentration of CO_2 is usually greater in subsoil probably due to more sluggish aeration in lower layer than in the surface soil.

2. Type of crop: Plant roots require oxygen, which they take from the soil air and deplete the concentration of oxygen in the soil air. Soils on which crops are grown contain more CO_2 than fallow lands. The amount of CO_2 is usually much greater near the roots of plants than further away. It may be due to respiration by roots.

3. Microbial activity: The microorganisms in soil require oxygen for respiration and they take it from the soil air and thus deplete its concentration in the soil air. Decomposition of organic matter produces CO_2 because of increased microbial activity. Hence, soils rich in organic matter contain higher percentage of CO_2.

4. Seasonal variation: The quantity of oxygen is usually higher in dry season than during the monsoon. Because soils are normally drier during the summer months, opportunity for gaseous exchange is greater during this period. This results in relatively high O_2 and low CO_2 levels. Temperature also influences the CO_2 content in the soil air. High temperature during summer season encourages microorganism activity which results in higher production of CO_2.

Exchange of Gases between Soil and Atmosphere

The exchange of gases between the soil and the atmosphere is facilitated by two mechanisms:

1. Mass flow: With every rain or irrigation, a part of the soil air moves out into the atmosphere as it is displaced by the incoming water. As and when moisture is lost by evaporation and transpiration, the atmospheric air enters the soil pores. The variations in soil temperature cause changes in the temperature of soil air. As the soil air gets heated during the day, it expands and the expanded air moves out into the atmosphere. On the other hand, when the soil begins to cool, the soil air contracts and the atmospheric air is drawn in.

2. Diffusion: Most of the gaseous interchange in soils occurs by diffusion. Atmospheric and soil air contains a number of gases such as nitrogen, oxygen, carbon dioxide etc., each of which exerts its own partial pressure in proportion to its concentration.

The movement of each gas is regulated by the partial pressure under which it exists. If the partial pressure on one of the gases is greater in the soil air than in the atmospheric air, it (CO_2) moves out into the atmosphere. Hence, the concentration of CO_2 is more in soil air.

On the other hand, partial pressure of oxygen is low in the soil air, as oxygen present in soil air is consumed as a result of biological activities. The oxygen present in the atmospheric air therefore, diffuses into the soil air till equilibrium is established. Thus, diffusion allows extensive movement and continual change of gases between the soil air and the atmospheric air. Oxygen and carbon dioxide are the two important gases that take in diffusion.

Importance of Soil Aeration

1. Plant and root growth: Soil aeration is an important factor in the normal growth of plants. The supply of oxygen to roots in adequate quantities and the removal of CO_2 from the soil atmosphere are very essential for healthy plant growth.

When the supply of oxygen is inadequate, the plant growth either retards or ceases completely as the accumulated CO_2 hampers the growth of plant roots. The abnormal effect of insufficient aeration on root development is most noticeable on the root crops. Abnormally shaped roots of these plants are common on the compact and poorly aerated soils. The penetration and development of root are poor. Such undeveloped root system cannot absorb sufficient moisture and nutrients from the soil.

2. Microorganism population and activity: The microorganisms living in the soil also require oxygen for respiration and metabolism. Some of the important microbial activities such as the decomposition of organic matter, nitrification, Sulphur oxidation etc depend upon oxygen present in the soil air. The deficiency of air in soil slows down the rate of microbial activity.

For example, the decomposition of organic matter is retarded and nitrification arrested. The microorganism population is also drastically affected by poor aeration.

3. Formation of toxic material: Poor aeration results in the development of toxin and other injurious substances such as ferrous oxide, H_2S gas, CO_2 gas etc in the soil.

4. Water and nutrient absorption: A deficiency of oxygen has been found to check the nutrient and water absorption by plants. The energy of respiration is utilized in absorption of water and nutrients. Under poor aeration condition, plants exhibit water and nutrient deficiency.

5. Development of plant diseases: Insufficient aeration of the soil also lead to the development of diseases. For example, wilt of gram and dieback of citrus and peach.

Influence of Soil Temperature and Air On Plant Growth

Effect of soil temperature on plant growth

a) Soil temperature requirements of plants: The soil temperature requirements of plants vary with the species. The temperature at which a plant thrives and produces best growth is called optimum range. The entire range of temperature under which a plant can grow including the optimum range is called growth range. The maximum and minimum temperatures beyond which the plant will die are called survival limits.

Range	Maize (°C)	Wheat (°C)
Optimum range	25 – 35	15 – 27
Growth range	10 – 39	5 - 35
Survival limits	0 – 43	0 - 43

b) Availability of soil water and plant nutrients: The free energy of water increases with temperature. Up to wilting point limit, warming of soil increases water availability beyond which it decreases. Low temperatures reduce the nutrient availability, microbial activities and root growth and branching. The ability to absorb nutrients and water by plants reduces at low temperatures.

Soil Temperature Management

Use of organic and synthetic mulches: Mulches keep soil cooler in hot summer and warm in cool winter.

Soil water management: High moisture content in humid temperate region lowers soil temperature.

Tillage management: Tilling soil to break the natural structure reduces the heat conductance and heat loss. A highly compact soil looses heat faster than loose friable soil.

Methods of measuring soil temperature: Mercury soil thermometers of different lengths, shapes and sizes with protective cover are buried at different depths to measure the temperature. Thermo couple and thermister based devices are also available. Infra-red thermo meters measure the surface soil temperature. Automatic continuous soil thermographs record the soil temperatures on a time scale. The International Meteorological Organization recommends standard depths to measure soil temperatures at 10, 20, 50 and 100 cm.

Soil Color

The first impression we have when looking at bare earth or soil is of color. Bright colors especially, catch our eye. Geographers are familiar with Red Desert soils in California, Arizona, and Nevada (Arizona State Soil); and Gray Desert soils in Idaho, Utah, and Nevada (Nevada State Soil). We have the White Sands in New Mexico, Green Sands along the Atlantic Coast, and Redbeds in Texas and Oklahoma (Oklahoma State Soil). The Red River between Oklahoma and Texas carries red sediment downstream, particularly in times of flood. The Yellow River (Hwang Ho) in China carries yellow sediment. Surface soils in the Great Plains and Corn Belt are darkened and enriched by organic matter.

Earth materials found in such locations as those mentioned above were used as coloring agents early in the development of most human cultures. As earth material was fashioned into utilitarian vessels, artistic colors inevitably were incorporated into them. Indigenous North American cultures used contrasting earth colors as body paints, and modern American culture uses colored earth in cosmetics and ceramics and as pigments for paints.

Munsell Color System

Red, brown, yellow, yellowish-red, grayish-brown, and pale red are all good descriptive colors of soil, but not very exact. Just as paint stores have pages of color chips, soil scientists use a book of color chips that follow the Munsell System of Color Notation. The Munsell System allows for direct comparison of soils anywhere in the world. The system has three components: hue (a specific color), value (lightness and darkness), and chroma (color intensity) that are arranged in books of color chips. Soil is held next to the chips to find a visual match and assigned the corresponding Munsell notation. For example, a brown soil may be noted as: hue value/chroma (10YR 5/3). With a soil color book with Munsell notations, a science student or teacher can visually connect soil colors with natural environments of the area, and students can learn to read and record the color, scientifically. Soil color by Munsell notation is one of many standard methods used to describe soils for soil survey. Munsell color notations can be used to define an archeological site or to make comparisons in a criminal investigation. Even carpet manufacturers use Munsell soil colors to match carpet colors to local soils so that the carpet will not show the dirt (soil) tracked into the house.

Soil Composition and Color

Soil color and other properties including texture, structure, and consistence are used to distinguish and identify soil horizons (layers) and to group soils according to the soil classification system called Soil Taxonomy. Color development and distribution of color within a soil profile are part of weathering. As rocks containing iron or manganese weather, the elements oxidize. Iron forms small crystals with a yellow or red color, organic matter decomposes into black humus, and manganese forms black mineral deposits. These pigments paint the soil (Michigan State Soil). Color is also affected by the environment: aerobic environments produce sweeping vistas of uniform or subtly changing color, and anaerobic (lacking oxygen), wet environments disrupt color flow with complex, often intriguing patterns and points of accent. With depth below the soil surface, colors usually become lighter, yellower, or redder.

Interpreting Soil Color

Color can be used as a clue to mineral content of a soil. Iron minerals, by far, provide the most and the greatest variety of pigments in earth and soil (see the following table).

Properties of Minerals

Mineral	Formula	Size	Munsell	Color
goethite	$FeOOH$	(1-2 m m)	10YR 8/6	yellow
goethite	$FeOOH$	(~0.2 m m)	7.5YR 5/6	strong brown
hematite	Fe_2O_3	(~0.4 m m)	5R 3/6	red
hematite	Fe_2O_3	(~0.1 m m)	10R 4/8	red
lepidocrocite	$FeOOH$	(~0.5 m m)	5YR 6/8	reddish-yellow
lepidocrocite	$FeOOH$	(~0.1 m m)	2.5YR 4/6	red
ferrihydrite	$Fe(OH)_3$		2.5YR 3/6	dark red
glauconite	$K(Si_xAl_4-x)(Al,Fe,Mg)O_{10}(OH)_2$		5Y 5/1	dark gray
iron sulfide	FeS		10YR 2/1	black
pyrite	FeS_2		10YR 2/1	black (metallic)
jarosite	$K Fe_3 (OH)_6 (SO_4)2$		5Y 6/4	pale yellow
todorokite	MnO_4		10YR 2/1	black
humus			10YR 2/1	black
calcite	$CaCO_3$		10YR 8/2	white
dolomite	$CaMg(CO_3)2$		10YR 8/2	white
gypsum	$CaSO_4 \times 2H_2O$		10YR 8/3	very pale brown
quartz	SiO_2		10YR 6/1	light gray

Relatively large crystals of goethite give the ubiquitous yellow pigment of aerobic soils. Smaller goethite crystals produce shades of brown. Hematite (Greek for blood-like) adds rich red tints. Large hematite crystals give a purplish-red color to geologic sediments that, in a soil, may be inherited from the geologic parent material. In general, goethite soil colors occur more frequently in temperate climates, and hematite colors are more prevalent in hot deserts and tropical climates.

Color - or lack of color - can also tell us something about the environment. Anaerobic environments occur when a soil has a high water table or water settles above an impermeable layer. In many soils, the water table rises in the rainy season. When standing water covers soil, any oxygen in the water is used rapidly, and then the aerobic bacteria go dormant. Anaerobic bacteria use ferric iron (Fe3+) in goethite and hematite as an electron acceptor in their metabolism. In the process, iron is reduced to colorless, water-soluble ferrous iron (Fe2+), which is returned to the soil.

Other anaerobic bacteria use Mn^{4+} as an electron acceptor, which is reduced to colorless, soluble Mn^{2+}. The loss of pigment leaves gray colors of the underlying mineral. If water stays high for long periods, the entire zone turns gray.

When the water table edges down in the dry season, oxygen reenters. Soluble iron oxidizes into characteristic orange colored mottles of lepidocrocite (same formula as goethite but different crystal structure) on cracks in the soil. If the soil aerates rapidly, bright red mottles of ferrihydrite form in pores and on cracks. Usually ferrihydrite is not stable and, in time, alters to lepidocrocite.

Along seacoasts, tide waters saturate soils twice daily, bringing soluble sulfate anions. Anaerobic bacteria use the sulfate as an electron acceptor and release sulfide (S^{2-}) which combines with ferrous iron to precipitate black iron sulfide. A little hydrochloric acid (HCl) dropped on this black pigment quickly produces a rotten egg odor of hydrogen sulfide (H_2S) gas. Soils that release H_2S gas are called sulfidic soils. With time, iron sulfide alters to pyrite (FeS_2) and imparts a metallic bluish color. If sulfidic soils are drained and aerated, they quickly become very acid (pH 2.5 to 3.5), and a distinctive pale yellow pigment of jarosite forms. This is the mark of an acid sulfate soil that is quite corrosive and grows few plants.

Galuconitic green sands form in shallow ocean water near a coast. They become part of soils that form after sea level drops. White colors of uncoated calcite, dolomite, and gypsum are common in geologic materials and soils in arid climates. A little carbonate dissolves in water, moves downward, and precipitates in soft white bodies or harder nodules. It also accumulates in root pores as lacy, dendritic (tree-branch) patterns.

Influence of Organic Matter on Soil Color

Soil has living organisms and dead organic matter, which decomposes into black humus. In grassland (prairie) soils the dark color permeates through the surface layers bringing with it nutrients and high fertility (Kansas State Soil). Deeper in the soil, the organic pigment coats surfaces of soil, making them darker than the color inside. Humus color decreases with depth and iron pigments become more apparent. In forested areas, organic matter (leaves, needles, pine cones, dead animals) accumulates on top of the soil. Water-soluble carbon moves down through the soil and scavenges bits of humus and iron that accumulate below in black, humic bands over reddish iron bands. Often, a white layer, mostly quartz occurs between organic matter on the surface where pigments were removed (Wisconsin State Soil).

Organic matter plays an indirect, but crucial role in the removal of iron and manganese pigments in wet soils. All bacteria, including those that reduce iron and manganese, must have a food source. Therefore, anaerobic bacteria thrive in concentrations of organic matter, particularly in dead roots. Here, concentrations of gray mottles develop.

Soil color is a study of various chemical processes acting on soil. These processes include the weathering of geologic material, the chemistry of oxidation-reduction actions upon the various minerals of soil, especially iron and manganese, and the biochemistry of the decomposition of organic matter. Other aspects of Earth science such as climate, physical geography, and geology all influence the rates and conditions under which these chemical reactions occur.

Soil Air

Soil air is a continuation of the atmospheric air. Unlike the other components, it is the constant state of motion from the soil pores into the atmosphere.

The amount of air or soil-air content is directly related to the bulk density of the soil and the amount of water in the soil profile. The bulk density of natural soil varies from approximately 1.0 Mg m−3 to 1.7–1.8 Mg m−3. Thus, the relative amount of void or pore space in the soil varies between approximately 30 and 60%. The soil pores or voids can be filled with either air or water. Therefore the soil-air content or air-filled porosity can vary between approximately 30 and 60%.

The composition of soil air depends on the relative magnitude of both the sources and the sinks of the various gas components, the interchange between soil air and atmospheric air, and the partitioning of the gases between the gaseous, liquid, and solid (mineral and organic matter) phases of the soil. If a soil were completely 'aerated' the concentrations of the gases in the soil airwould be similar to that in the atmosphere. Oxygen concentrations in the soil air will be somewhat below that in the atmosphere (approximately 20% by volume), since O_2 is consumed in soil by plant root and microbial respiration and through chemical reactions. Under some conditions, O_2 concentrations can fall to zero and the soil becomes anaerobic (anoxic). It is now widely accepted that under some conditions soil profiles do not have to be either fully aerated or fully anaerobic but may be partially aerobic and partially anaerobic. Anaerobic pockets or 'hot spots' may exist within the soil due to pockets of very high O_2 consumption such as around incorporated carbon materials and/or due to very slow diffusion to regions of O_2 consumption. For example, the interior of large aggregates may be anaerobic for these reasons.

CO_2 concentrations in the soil air can be as high as 10 times more than in the atmosphere (0.036% by volume). Since nitrogen gas (N_2) is more abundant than other gases in the atmosphere (approx. 78%) and there are generally no sources or sinks for N_2 in the soil (except N_2 absorbed during nitrogen fixation or produced during denitrification), the concentration of N_2 in the soil air will be similar to that in the atmosphere, varying only slightly depending on the production and consumption of other soil gases. The soil air will also contain varying amounts of nitric oxide and nitrous oxide (from nitrification and denitrification); methane, hydrogen sulfide, and ethylene (from anaerobic processes); water vapor; and trace amounts of inert gases such as argon. Human activities also result in the accidental or intentional introduction of gases in the soil profile such as fumigants, anhydrous ammonia, pesticides, and various volatile organic chemicals that exist partially in the vapor phase.

Soil Water

The capacity of soil to regulate the terrestrial freshwater supply is a fundamental ecosystem service. Water percolating through soil is filtered, stored for plant utilization, and redistributed across flow paths to groundwater and surface water bodies. As such, the sustainability of water resources (considering both quantity and quality) is directly influenced by soil. Thus, most aspects of terrestrial- and freshwater aquatic-life depend on hydrologic processes in soil.

Water dynamics in soil are governed by many factors that change vertically with depth, laterally across landforms and temporally in response to climate.

Storage, Flow, and Potential Energy

Stored water in soil is a dynamic property that changes spatially in response to climate, topography and soil properties, and temporally as a result of differences between utilization and redistribution via subsurface flow. Changes in soil moisture storage can be generalized with a mass balance equation (eq. 1) as a result of the difference between the amount of water added and that which is lost.

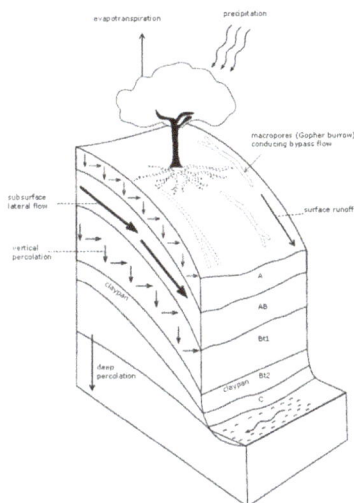

Figure: Conceptual diagram of a soil profile illustrating the multiple flow paths through which water moves through soil.

Change in soil moisture storage = inputs − outputs

Water content increases (positive change in storage) when inputs including precipitation or irrigation exceed outputs. Water content decreases (negative change in storage) when outputs such as deep percolation, surface runoff, subsurface lateral flow, and evapotranspiration (ET) exceed inputs.

Water storage and redistribution are a function of soil pore space and pore-size distribution, which are governed by texture and structure. Generally speaking, clay-rich soils have the largest pore space, hence the greatest total water holding capacity. However, total water holding capacity does not describe how much water is available to plants, or how freely water drains in soil. These processes are governed by potential energy. Water is stored and redistributed within soil in response to differences in potential energy. A potential energy gradient dictates soil moisture redistribution and losses, where water moves from areas of high- to low-potential energy. When at or near saturation, soils typically display water potentials near 0 MPa. Negative water potentials arise as soil dries resulting in suction or tension on water allowing the soil to retain water like a sponge.

Three soil moisture states, saturation, field capacity and permanent wilting point are used to describe water content across different water potentials in soil and are related to the energy required to move water (or extract water from soil). When the soil is at or near saturation the direction of the potential energy gradient is downward through the soil profile or laterally down slope. This mechanism of flow by the force of gravity occurs mainly in macropores. As the soil dries, field capacity is reached after free drainage of macropores has occurred. Field capacity represents the soil water content retained against the force of gravity by matric forces (in micropores and mesopores) at tension of -0.033 MPa. As water content decreases, soil matric potential decreases, becoming more negative, and as a result, water is held more strongly to mineral surfaces due to cohesive forces between water molecules and adhesive forces associated with water and mineral particles (capillary forces). Water held between saturation and field capacity is transitory, subject to free drainage over short time periods, hence is it is generally considered unavailable to plants. This free water is termed drainable porosity. In contrast, much of the water held at field capacity is available for plant uptake and use through evapotranspiration.

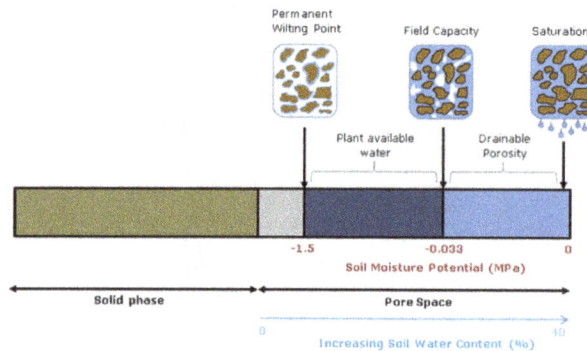

Figure: Water content and water potential at saturation,
field capacity and permanent
wilting point.

The difference in water content between field capacity and permanent wilting point is plant available water. Drainable porosity is the amount of water that drains from macropores by gravity between saturation to field capacity typically representing three days of drainage in the field.

The point at which matric forces hold water too tightly for plant extraction (-1.5 MPa) is termed the permanent wilting point. The amount of water held between field capacity and permanent wilting point is considered plant available water (PAW). Water held between these two states is retained against the force of gravity, but not so tightly that it cannot be extracted by plants. Mesopores and micropores supply most plant available water. Water held at potentials below permanent wilting point (< -1.5 MPa) is not available for use by most plants because it strongly adheres to mineral particles. Water held at permanent wilting point is associated with partially filled micropores and hydrated surfaces of soil particles.

Influence of Texture and Structure

Differences in soil properties (texture and structure) affect the water content at saturation, field capacity, and permanent wilting point. Texture and structure determine pore size distribution in soil, and therefore, the amount of PAW. Figure illustrates how the magnitude of PAW changes with soil texture. Coarse textured soils (sands and loamy sands) have low PAW because the pore size distribution consists mainly of large pores with limited ability to retain water. Although fine textured soils have the highest total water storage capacity due to large porosity values, a significant fraction of water is held too strongly (strong matric forces/low, negative water potentials) for plant uptake. Fine textured soils (clays, sandy clays and silty clays) have moderate PAW because their pore size distribution consists mainly of micropores. Loamy textured soils (loams, sandy loams, silt loams, silts, clay loams, sandy clay loams and silty clay loams) have the highest PAW, because these textural classes give rise to a wide range in pore size distribution that results in an ideal combination of meso- and micro-porosity. Soil structure can increase PAW by increasing porosity. Soil depth and rock fragment content also affect water holding capacity because bedrock and rock fragments are assumed to be unable to hold plant available water and/or accommodate plant roots.

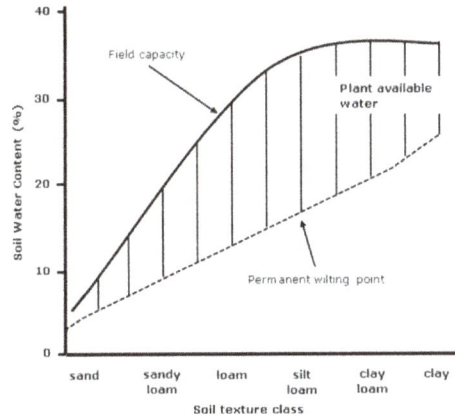

Figure: Generalized relationship between soil texture classes and plant available water holding capacity.

Water movement in soil is closely linked with storage because water potential is a function of water content. Water flow is also influenced by texture and structure, and other factors such as the layering of soil profiles. The rate of water flow is a function of the potential energy gradient and the ease with which water is transmitted through soil, termed saturated hydraulic conductivity, which is governed by pore size distribution and tortuosity of flow paths. Clay-rich soils have low saturated hydraulic conductivity due to a highly tortuous flow path. Conversely, sandy soils have larger pores and lower tortuosity that facilitate rapid water flow. The National Cooperative Soil Survey identifies the following permeability classes based on soil texture (Table). These classes can be modified (qualitatively) by the degree of soil structure. Strong soil structure, consisting of very fine and fine aggregates (e.g., granular, fine and medium angular blocky and sub angular blocky) facilitate rapid drainage of soil by increasing macro porosity. In contrast, weak structure or coarse-sized structural units (prismatic or blocky) and platy structure can inhibit flow, creating a more tortuous flow path constraining water to inter-structural voids. Soil structure is highly relevant to water management in soils because it is subject to change either through deterioration by improper management, or to improvement through additions of soil organic matter. In contrast, it is usually infeasible to change texture.

Permeability Class	Permeability (cm/hr)	Textural class
Very slow	<0.13	clay
Slow	0.13–0.5	sandy clay, silty clay
Moderately slow	0.05–2.0	clay loam, sandy clay loam, silty clay loam
Moderate	2.0–6.3	very fine sandy loam, loam, silt loam, silty clay loam, silt
Moderately rapid	6.3–12.7	sandy loam, fine sandy loam
Rapid	12.7–25.4	sand, loamy sand
Very Rapid	>25.4	coarse sand

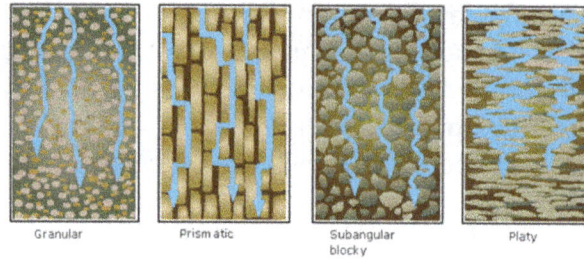

Figure: Water movement through different soil structure shapes.

Drainage

The ease with which water drains from soil is equally as important as storage. For example, most terrestrial plants need to assimilate oxygen through roots, but oxygen is scarce in saturated soils. Moreover, microbial decomposition of organic matter is greatest (by orders of magnitude) under aerobic conditions. Poorly drained soils have limitations for a variety of land use practices. The recognition of poor drainage in soils is also used in wetland delineation efforts.

Drainage capacity can be identified through careful observation of soil properties. Poorly drained soils result in episodes of prolonged saturation, while excessively drained soils commonly experience water deficits. Soils that are saturated at times when the system is above biological zero (> 5° C) often develop redoximorphic features. The most common redoximorphic features in soil are iron and manganese concentrations and iron depletions. These features arise through microbial decomposition of soil organic matter under anaerobic conditions. Anaerobic conditions arise because the diffusion of oxygen in saturated soil is very slow and does not keep up with oxygen demands of aerobic respiration by microbes. When oxygen is depleted, facultative microbes utilize iron (Fe^{3+}) and manganese (Mn^{4+}) as terminal electron acceptors to make energy. In doing so, these elements are reduced and become soluble in soil solution. As water mobilizes these soluble constituents they eventually encounter air (e.g., in root channels or other macropores) where they oxidize and re-precipitate as iron and manganese concentrations. Iron concentrations are usually rust colored, red or orange. Manganese concentrations are gunmetal blue, almost black. Redox depletions arise in microsites of extreme reducing conditions where Fe^{3+} has been completely removed. Redox depletions are typically dull in color (low Munsell chroma usually 2 or less for taxonomic purposes) and have blue, green, pale brown or yellow hues that reflect the coloration of primary minerals. The depth at which redoximorphic features occur is used to describe the extent of saturated conditions within a soil profile.

Figure: Iron redox concentrations in the form of rusty orange pore linings.

Figure: Gunmetal blue manganese redox concentrations in the form of ped surface coatings and pore linings. Note 35 mm lens cap at base of photo for scale.

Figure: Example of pale bluish gray redox depletions. Note the faint rusty orange concentration distributed throughout the soil matrix.

The National Cooperative Soil Survey describes five soil drainage classes determined by the rate at which water drains from soil and the height of the water table during the growing season (Table). Drainage classes are used for a variety of land-use decisions such as soil suitability for groundwater banking, land application of wastes, engineering and construction, septic system development, crop selection, land capability classification and wetland habitat.

Drainage Class	Water Removed	Redoximorphic Features	Water Table Height (m)
Excessively drained	Very rapid	None	>1.5
Somewhat excessively drained	Rapid	None	>1.5
Well drained	Readily not rapidly	Below 1 m	1.0–1.5
Moderately well drained	Somewhat slowly	Below 0.5 m	0.5–1.0
Somewhat poorly drained	Slowly	Below 0.25 m	0.25–0.5
Poorly drained	Very slowly	Gleyed or at or near soil surface	<0.25
Very poorly drained	Persistent saturation throughout the growing season		

Timing of Inputs

The amount and timing of precipitation ultimately governs soil moisture content, availability and flow. The temporal nature of moisture dynamics dictates ecosystem response and land-use decisions. Soil moisture regimes are used in Soil Taxonomy to describe annual variability in moisture as dictated by climatic factors and indirectly by soil and landscape factors. There are five general soil moisture regimes: aquic, udic, xeric, ustic, and aridic. These moisture regimes have detailed definitions.

Aquic - poorly drained soils that are saturated when the soil temperature (at 50 cm) is above biological zero (>5°C). These soils typically display evidence of prolonged saturation in the form of redoximorphic features in the root zone.

Udic - soils typical of humid or tropical environments that receive an even distribution of precipitation throughout the year. Soils are never dry for as long as 90 cumulative days or 60 consecutive days following summer solstice.

Ustic - soil moisture conditions intermediate between aridic and udic and typical of semiarid, tropical and monsoon climates, where soil is moist during part or all of the growing season followed by a prolonged dry season at some point in the year.

Xeric - a soil moisture regime with wet winters and dry summers typical of Mediterranean type climates.

Aridic - a soil moisture regime that is dry (low plant available water) for most times of the year.

Figure depicts these soil moisture regimes across the globe. Soil moisture regimes serve as a broad scale planning tool to inform a variety land use decisions related to the hydrologic cycle such as, regions suitable for groundwater banking, dry land agriculture, summer fallow, or requiring irrigation technology, engineering considerations, and drainage infrastructure. These decisions cannot be made from climatic conditions alone. Soil moisture regimes are more useful in this type of decision making because it accounts for how soil properties (such as PAW) affect moisture dynamics within the soil profile.

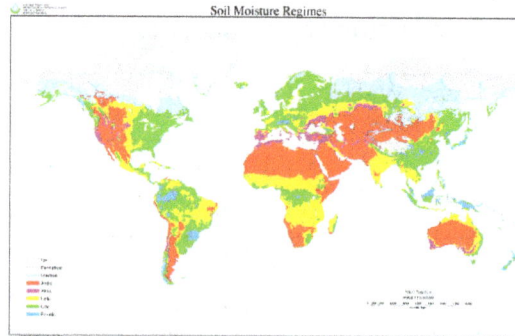

Figure: Global map of soil moisture regimes. Developed by USDA-NRCS, Soil Survey Division, World Soil Resources, Washington, DC.

Integrative Case Study

The type of soil moisture regime can be used to infer the direction of water movement in soil, which affects soil development and morphology. The following case study illustrates this point. Figure shows two soils, Mollisols (Xerolls and Ustolls) with 450 mm of mean annual precipitation that differ in the timing of precipitation (i.e., the soil moisture regime). Mollisols are typically described as grassland soils that have large stocks of below ground soil organic matter. The Ustoll is a Mollisol with an ustic soil moisture regime that formed in a bunchgrass prairie east of the Rocky Mountains. The Xeroll is a Mollisol with a xeric soil moisture regime that formed in a bunchgrass prairie on the Columbia Plateau in eastern Washington. These soils share similarities in soil forming factors including parent material, vegetation, age, and topography. The notable difference between these two profiles is the nature of the B horizons (subsurface zones of accumulation of clay, salts or iron). The Ustoll has a Bk horizon, which is a zone of accumulation of calcium carbonate (CaCO3). This semi-soluble salt tends to accumulate in sub soils where deep percolation is minimal. The Xeroll has a Bw horizon, subsoil where salts have been removed by deep percolation and the development of soil structure and slight accumulation of clay are detectable.

Figure: Examples of Mollisols (grassland soils) formed from loess. Both soils have mean annual precipitation of 450 mm. The Ustoll (a) has an ustic soil moisture regime and the Xeroll (b) has a xeric soil moisture regime. Note the presence of calcium carbonate filaments and masses throughout the Bk horizon of the Ustoll (a).

Potential evapotranspiration and precipitation over time in these soils illustrate the effects of timing of rainfall on soil moisture dynamics. In the Xeroll, precipitation is low during the growing season facilitating a period of utilization of stored water during the early growing season followed by a prolonged deficit. Soil moisture is recharged at the onset of fall rains when the dormant season begins. As precipitation continues throughout the winter months a surplus is achieved. Surplus occurs when soils become saturated and water is allowed to freely drain with the force of gravity. A surplus in soil moisture results in the loss of free water stored between saturation and field capacity, which is subject to gravitational flow. In this scenario there are two main flow paths: 1) deep percolation of free water beyond the root zone, and 2) losses of stored water by ET during the growing season. The large proportion of surplus water present during the winter months when plants are dormant facilitates the leaching of soluble and semi-soluble salts out of the soil profile and into the groundwater.

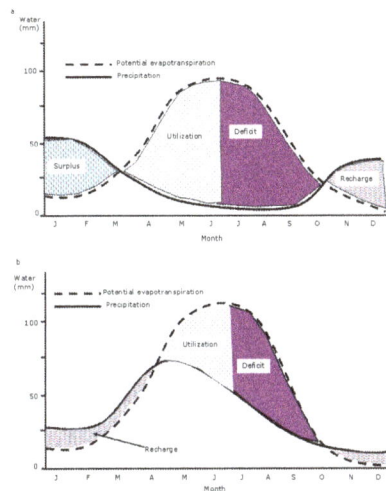

Figure: Climatic information and soil water balance for an Ustoll (a) and Xeroll (b).
Note that the Ustoll never reaches a surplus condition, free drainage of saturated macropores.

In the Ustoll, most precipitation occurs during the growing season and is utilized through transpiration until a soil moisture deficit occurs. As evapotranspiration decreases during the dormant season, soil recharge occurs, but precipitation is not high enough at this time to create a surplus. As a result, soil moisture is stored over the dormant season (winter months) and the major flow path is upward via capillary rise as plants become active in spring, extracting water by evapotranspiration. The absence of a surplus results in incomplete leaching of the soil profile and the accumulation of semi soluble salts (white coatings), as symbolized by the Bk horizon.

This comparison of ustic and xeric soil moisture regimes provides an example of how soil moisture dynamics and timing of precipitation govern water utilization and movement in soil. Timing of precipitation is equally as important as its magnitude when considering soil water dynamics. This case study demonstrates that the fate of water in soil (e.g., deep percolation or evapotranspiration) governs the expression of soil morphologic features. Thus, soil features can be used to infer hydrologic processes.

Water storage dynamics and flow facilitate the four basic soil forming processes: translocations, transformations, additions and losses of soil constituents in a soil profile. These processes determine the chemical, morphological and physical properties of soil such as the variation of texture with depth. Hydrological processes active in soil contribute to weathering processes, and indicators of these processes are preserved by the soil profile in the form of observable and measurable soil characteristics, similar to those discussed in the case study. Other soil morphologic indicators of hydrologic process include redoximorphic features, abrupt accumulation of clay in the subsoil, development of soil structure, and presence of cemented layers. Thus soil resource inventories (e.g., soil survey), which document soil properties, can be used to infer hydrologic processes.

Although climatic factors ultimately control plant-water relationships, soils regulate water acting as a sponge to hold water against gravitational forces in plant available form. Soil properties such as texture and structure govern pore size distribution, which dictates total water storage, available water holding capacity, and water movement in soil. While it is generally not feasible to modify soil texture to improve plant-water relationships, soil structure can be enhanced by adding organic matter to promote more meso- and macro-porosity, which increases plant available water holding capacity and helps to promote free drainage. An understanding of soil water relationships is fundamental to most land use decisions.

Soil Salinity

The beginning of 21st century is marked by global scarcity of water resources, environmental pollution and increased salinization of soil and water. Increasing human population and reduction in land available for cultivation are two threats for agricultural sustainability. Various environmental stresses viz. high winds, extreme temperatures, soil salinity, drought and flood have affected the production and cultivation of agricultural crops, among this soil salinity is one of the most devastating environmental stresses, which causes major reductions in cultivated land area, crop productivity and quality. A saline soil is generally defined as one in which the electrical conductivity (EC) of the saturation extract (ECe) in the root zone exceeds 4 dS m−1 (approximately 40 mM NaCl) at 25 °C and has an exchangeable sodium of 15%. The yield of most crop plants is reduced at this ECe, though many crops exhibit yield reduction at lower ECes. It has been estimated that worldwide 20% of total cultivated and 33% of irrigated agricultural lands are afflicted by high salinity. Furthermore, the salinized areas are increasing at a rate of 10% annually for various reasons, including low precipitation, high surface evaporation, weathering of native rocks, irrigation with saline water, and poor cultural practices. It has been estimated that more than 50% of the arable land would be salinized by the year 2050.

Water and soil management practices have facilitated agricultural production on soil marginalized by salinity but an additional gain from these approaches seems problematic. Impacted soils are a major limiting production factor worldwide for every major crop. A significant increase (an estimated 50%) in grain yields of major crop plants such as rice, wheat and maize is required to fulfill the food supply requirements for the projected population by 2050. The urgency of feeding the world's growing population while combating soil pollution, salinization, and desertification has given plant and soil productivity research vital importance. Under such circumstances, it requires suitable biotechnology not only to improve crop productivity but also to improve soil health through interactions of plant roots and soil microorganisms.

Salt stressed soils are known to suppress the growth of plants. Plants in their natural environment are colonized both by endocellular and intracellular microorganisms. Rhizosphere microorganisms, particularly beneficial bacteria and fungi, can improve plant performance under stress environments and, consequently, enhance yield both directly and indirectly. Some plant growth-promoting rhizobacteria (PGPR) may exert a direct stimulation on plant growth and development by providing plants with fixed nitrogen, phytohormones, iron that has been sequestered by bacterial siderophores, and soluble phosphate. Others do this indirectly by protecting the plant against soil-borne diseases, most of which are caused by pathogenic fungi. The problem of soil salinization is a scourge for agricultural productivity worldwide. Crops grown on saline soils suffer on an account of high osmotic stress, nutritional disorders and toxicities, poor soil physical conditions and reduced crop productivity.

Problem of Soil Salinization

Soil salinity is an enormous problem for agriculture under irrigation. In the hot and dry regions of the world the soils are frequently saline with low agricultural potential. In these areas most crops are grown under irrigation, and to exacerbate the problem, inadequate irrigation management leads to secondary salinization that affects 20% of irrigated land worldwide. Irrigated agriculture is a major human activity, which often leads to secondary salinization of land and water resources in arid and semi-arid conditions. Salts in the soil occur as ions (electrically charged forms of atoms or compounds). Ions are released from weathering minerals in the soil. They may also be applied through irrigation water or as fertilizers, or sometimes migrate upward in the soil from shallow groundwater. When precipitation is insufficient to leach ions from the soil profile, salts accumulate in the soil resulting soil salinity. All soils contain some water-soluble salts. Plants absorb essential nutrients in the form of soluble salts, but excessive accumulation strongly suppresses the plant growth. During the last century, physical, chemical and/or biological land degradation processes have resulted in serious consequences to global natural resources (e.g. compaction, inorganic/organic contamination, and diminished microbial activity/diversity). The area under the affected soils continues to increase each year due to introduction of irrigation in new areas.

Salinization is recognized as the main threats to environmental resources and human health in many countries, affecting almost 1 billion ha worldwide/globally representing about 7% of earth's continental extent, approximately 10 times the size of a country like Venezuela or 20 times the size of France.

Impact of Salinity oPlants

Agricultural crops exhibit a spectrum of responses under salt stress. Salinity not only decreases the agricultural production of most crops, but also, effects soil physicochemical properties, and ecological balance of the area. The impacts of salinity include—low agricultural productivity, low economic returns and soil erosions,. Salinity effects are the results of complex interactions among morphological, physiological, and biochemical processes including seed germination, plant growth, and water and nutrient uptake. Salinity affects almost all aspects of plant development including: germination, vegetative growth and reproductive development. Soil salinity imposes ion toxicity, osmotic stress, nutrient (N, Ca, K, P, Fe, Zn) deficiency and oxidative stress on plants, and thus limits water uptake from soil. Soil salinity significantly reduces plant phosphorus (P) uptake because phosphate ions precipitate with Ca ions. Some elements, such as sodium, chlorine, and boron, have specific toxic effects on plants. Excessive accumulation of sodium in cell walls can rapidly lead to osmotic stress and cell death. Plants sensitive to these elements may be affected at relatively low salt concentrations if the soil contains enough of the toxic element. Because many salts are also plant nutrients, high salt levels in the soil can upset the nutrient balance in the plant or interfere with the uptake of some nutrients. Salinity also affects photosynthesis mainly through a reduction in leaf area, chlorophyll content and stomatal conductance, and to a lesser extent through a decrease in photosystem II efficiency. Salinity adversely affects reproductive development by inhibiting microsporogenesis and stamen filament elongation, enhancing programed cell death in some tissue types, ovule abortion and senescence of fertilized embryos. The saline growth medium causes many adverse effects on plant growth, due to a low osmotic potential of soil solution (osmotic stress), specific ion effects (salt stress), nutritional imbalances, or a combination of these factors. All these factors cause adverse effects on plant growth and development at physiological and biochemical levels, and at the molecular level.

In order to assess the tolerance of plants to salinity stress, growth or survival of the plant is measured because it integrates the up- or down-regulation of many physiological mechanisms occurring within the plant. Osmotic balance is essential for plants growing in saline medium. Failure of this balance results in loss of turgidity, cell dehydration and ultimately, death of cells. On the other hand, adverse effects of salinity on plant growth may also result from impairment of the supply of photosynthetic assimilates or hormones to the growing tissues. Ion toxicity is the result of replacement of K^+ by Na^+ in biochemical reactions, and Na^+ and Cl^- induced conformational changes in proteins. For several enzymes, K^+ acts as cofactor and cannot be substituted by Na^+. High K^+ concentration is also required for binding tRNA to ribosomes and thus protein synthesis. Ion toxicity and osmotic stress cause metabolic imbalance, which in turn leads to oxidative stress. The adverse effects of salinity on plant development are more profound during the reproductive phase. Wheat plants stressed at 100–175 mM NaCl showed a significant reduction in spikelets per spike, delayed spike emergence and reduced fertility, which results in poor grain yields. However, Na^+ and Cl^- concentrations in the shoot apex of these wheat plants were below 50 and 30 mM, respectively, which is too low to limit metabolic reactions. Hence, the adverse effects of salinity may be attributed to the salt-stress effect on the cell cycle and differentiation. Salinity arrests the cell cycle transiently by reducing the expression and activity of cyclins and cyclin-dependent kinases that results in fewer cells in the meristem, thus limiting growth. The activity of cyclin-dependent kinase is diminished also by post-translational inhibition during salt stress. Recent reports also show that salinity adversely affects plant growth and development, hindering seed germination, seedling growth, enzyme activity, DNA, RNA, protein synthesis and mitosis.

Amelioration of Salinity

Salinization can be restricted by leaching of salt from root zone, changed farm management practices and use of salt tolerant plants. Irrigated agriculture can be sustained by better irrigation practices such as adoption of partial root zone drying methodology, and drip or micro-jet irrigation to optimize use of water. The spread of dry land salinity can be contained by reducing the amount of water passing beyond the roots. This can be done by re-introducing deep rooted perennial plants that continue to grow and use water during the seasons that do not support annual crop plants. This may restore the balance between rainfall and water use, thus preventing rising water tables and the movement of salt to the soil surface. Farming systems can change to incorporate perennials in rotation with annual crops (phase farming), in mixed plantings (alley farming, intercropping), or in site-specific plantings (precision farming). Although the use of these approaches to sustainable management can ameliorate yield reduction under salinity stress, implementation is often limited because of cost and availability of good water quality or water resource. Evolving efficient, low cost, easily adaptable methods for the abiotic stress management is a major challenge. Worldwide, extensive research is being carried out, to develop strategies to cope with abiotic stresses, through development of salt and drought tolerant varieties, shifting the crop calendars, resource management practices etc. as shown in Fig.

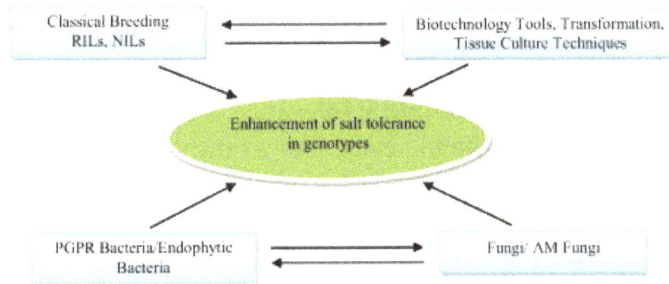

Figure: Different approaches for improvement of salt tolerance in agricultural crops.

Use of Salt Tolerant Crops and Transgenics

Using the salt-tolerant crops is one of the most important strategies to solve the problem of salinity. Tolerance will be required for the "de-watering" species, but also for the annual crops to follow, as salt will be left in the soil when the water table is lowered. Salt tolerance in crops will also allow the more effective use of poor quality irrigation water. To increase the plant salt-tolerance, there is a need for understanding the mechanisms of salt limitation on plant growth and the mechanism of salt tolerance at the whole-plant, organelle, and molecular levels. Under saline conditions, there is a change in the pattern of gene expression, and both qualitative and quantitative changes in protein synthesis. Although it is generally agreed that salt stress brings about quantitative changes in protein synthesis, there is some controversy as to whether salinity activates specialized genes that are involved in salt stress. Salt tolerance does not appear to be conferred by unique gene(s). When a plant is subjected to abiotic stress, a number of genes are turned on, resulting in increased levels of several metabolites and proteins, some of which may be responsible for conferring a certain degree of protection to these stresses. Efforts to improve crop performance by transgenic approach under environmental stresses have not been that fruitful because the fundamental mechanisms of stress tolerance in plants remain to be completely understood.

Development of salt-tolerant crops has been a major objective of plant breeding programs for decades in order to maintain crop productivity in semiarid and saline lands. Although several salt-tolerant varieties have been released, the overall progress of traditional breeding has been slow and has not been successful as only few major determinants genetic traits of salt tolerance have been identified. 25 years ago Epstein et al. described the technical and biological constraints to solving the problem of salinity. Although there has been some success with technical solutions to the problem, the biological solutions have been more difficult to develop because a pre-requisite for the development of salt tolerant crops is the identification of key genetic determinants of stress tolerance. The existence of salt-tolerant plants (halophytes) and differences in salt tolerance between genotypes within salt-sensitive plant species (glycophytes) indicates that there is a genetic basis to salt response. Although a lot of approaches have been done for development of salt tolerant plants by transgenics complete success is not achieved yet. The assessment of salt tolerance in transgenic experiments has been mostly carried out using a limited number of seedlings or mature plants in laboratory experiments. In most of the cases, the experiments were carried out in greenhouse conditions where the plants were not exposed to those conditions that prevail in high-salinity soils (e.g. alkaline soil pH, high diurnal temperatures, low humidity, and presence of other sodic salts and elevated concentrations of selenium and/or boron). The salt tolerance of the plants in the field needs to be evaluated and, more importantly, salt tolerance needs to be evaluated as a function of yield. The evaluation of field performance under salt stress is difficult because of the variability of salt levels in field conditions and the potential for interactions with other environmental factors, including soil fertility, temperature, light intensity and water loss due to transpiration. Evaluating tolerance is also made more complex because of variation in sensitivity to salt during the life cycle. For example, in rice, grain yield is much more affected by salinity than in vegetative growth. In tomato, the ability of the plants to germinate under conditions of high salinity is not always correlated with the ability of the plant to grow under salt stress because both are controlled by different mechanisms, although some genotypes might display similar tolerance at germination and during vegetative growth. Therefore, the assessment of stress tolerance in the laboratory often has little correlation to tolerance in the field. Although there have been many successes in developing stress-tolerant transgenics in model plants such as tobacco, Arabidopsis or rice, there is an urgent need to test these successes in other crops. There are several technical and financial challenges associated with transforming many of the crop plants, particularly the monocots. First, transformation of any monocot other than rice is still not routine and to develop a series of independent homozygous lines is costly, both in terms of money and time. Second, the stress tolerance screens will need to include a field component because many of the stress tolerance assays used by basic researchers involve using nutrient-rich media (which in some cases include sucrose). This type of screen is unlikely to have a relationship to field performance. Third, because saline soils are often complex and can include $NaCl$, $CaCl_2$, $CaSO_4$, Na_2SO_4, high boron concentrations and alkaline pH, plants that show particular promise will eventually have to be tested in all these environments.

An ideal sustainable agricultural system is one which maintains and improves human health, benefits producers and consumers both economically and spiritually, protects the environment, and produces enough food for an increasing world population. One of the most important constraints to agricultural production in world is abiotic stress conditions prevailing in the environment. Plant-associated microorganisms can play an important role in conferring resistance to abiotic stresses. These organisms could include rhizoplane, rhizosphere and endophytic bacteria and symbiotic fungi and operate through a variety of mechanisms like triggering osmotic response, providing growth hormones and nutrients, acting as bio control agents and induction of novel genes in plants. The development of stress tolerant crop varieties through genetic engineering and plant breeding is essential but a long drawn and expensive process, whereas microbial inoculation to alleviate stresses in plants could be a more cost effective environmental friendly option which could be available in a shorter time frame. Taking the current leads available, concerted future research is needed in this area, particularly on field evaluation and application of potential organisms as biofertilizers in stressed soil.

References

- Soil-structure-definition-types-and-formation-1129: soilmanagementindia.com, Retrieved 12 March 2018

- Soil-resistivity: circuitglobe.com, Retrieved 28 April 2018

- Soil-air, earth-and-planetary-sciences: sciencedirect.com, Retrieved 17 June 2018

- Soil-water-dynamics-59718900: nature.com, Retrieved 27 May 2018

- Pore-space-types-and-factors-affecting-it-soil-3489: soilmanagementindia.com, Retrieved 28 June 2018

Chapter 6

Soil Nutrient Cycle

The soil stores and controls the release of nutrients. This is referred to as the soil nutrient cycle. In order to completely understand the soil nutrient cycle, it is necessary to understand the processes related to it, such as ion absorption by plant roots, carbon cycle in soil, nitrogen cycle in soil, etc. which have been dealt in extensive detail in this chapter.

Ion Absorption by Plant Roots

Ion Absorption in Plants– Passive and Active Uptake

Passive Uptake of Ion Absorption

Non-Mediated Passive Uptake

Numerous investigators have demonstrated non-metabolic or passive uptake of ions due to the fact that when a plant cell or tissue is transferred from a low-salt concentration medium to a relatively high-salt concentration medium, there is an initial uptake of ions due to diffusion.

This initial rapid uptake is temperature independent and remains unaffected by the application of metabolic inhibitors. Passive absorption includes theories of diffusion, ion exchange, Donnan equilibrium, mass flow, etc.

(a) Diffusion:

If the above cell or tissue is again returned to a lower-salt solution, some of the absorbed ions will diffuse out into the external medium. This process of free diffusion of solutes follows the simple laws of diffusion. This free diffusion means free movement of ions in and out of the tissue.

The part of the tissue for free diffusion shows equilibrium with the external medium. This part of a cell or tissue is referred to as outer space. The term apparent free space is also used to denote the apparent volume accommodating the freely diffused ions.

(b) Ion Exchange:

In the ion exchange mechanism the ions within the cells are exchanged for the ions of equivalent charge of the external solution. If a solution of a dissociable salt (A+B−) is separated from distilled water by a membrane which is permeable to both ions diffusion will occur until the concentration of salt on both sides are equal. If the membrane is impermeable to either cation or anion, there will be no ion movement.

Now, if two solutions containing different salts ($A_1^+B_1^-$ and $A_2^+B_2^-$) are separated from one another by either a cation-permeable/anion-impermeable or cation-impermeable/anion- permeable membrane, one of the two ion species in each case is free to move across in exchange for an ion of same charge.

At equilibrium the ratio $[A_1^+]/[A_2^+]$ (when cation is permeable) or $[B_1^-]/[B_2^-]$ (when anion is permeable) is equal on the two sides.

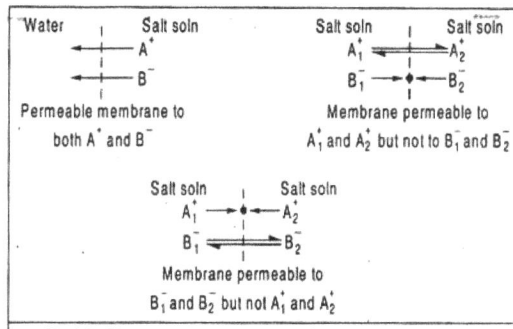

The total concentration of salts on either side is not affected by the exchanges of ions. Exchange involves equivalent electrical charge, so that two univalent ions are exchanged for one bivalent, three for one trivalent ion, and so on.

With the help of radioactive isotopes it is possible to calculate the rates of exchange of ions between the two solutions kept separated by an ion exchange membrane from measurements of changes in radioactivity with time on either side.

(c) Donnan Equilibrium:

The Donnan equilibrium theory accounts for the effect of fixed or non-diffusible ions and explains the co-operation of both electrical as well as diffusion phenomena. It is a complex ion-exchange system in which the membrane is impermeable to certain ions called fixed ions (Fig.).

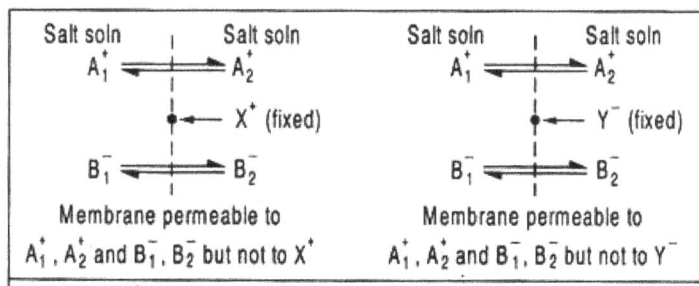

Fig. Diagrammatic representation of Donnan equilibrium

In both the cases represented in the figure X+ and Y– are fixed ions and cannot move from right to left. Due to the presence of these fixed ions some extra ions are absorbed against the concentration gradient. In the first case where X+ is fixed, equal numbers of cations and anions from the left-hand side will diffuse across the membrane until equilibrium is established. This equilibrium would also be electrically balanced.

However, additional anions are needed to balance the positive charges of the fixed cations on the right-hand side of the membrane. Therefore, the anion concentration would become greater on the right-hand side than it is on the left-hand side. Similarly, in the second case where Y− is fixed, cation accumulation takes place at equilibrium.

Thus, the accumulation of ions against a concentration gradient can occur without the participation of metabolic energy until a Donnan equilibrium is reached.

Since the concentrations of mobile ions are unequal on the two sides of the membrane in a Donnan system at equilibrium while the electrochemical potentials are unequal, it follows that there is an electrical potential difference between the two sides. This is sometimes called their Donnan membrane potential.

(d) Mass Flow of Ions:

Many workers believe that ions are absorbed by the roots along with the mass flow of water influenced by the transpiration stream. The theory states that, an increased transpiration rate causes an increase in absorption of ions. Lopushinsky (1964) working with de-topped tomato plants indirectly supported the concept that an increase in transpiration could increase the absorption of salts.

Some workers claim that transpiration pull indirectly affects ion absorption by continuous removal of ions after they have been released into the xylem ducts. Whatever may be the fact, mass flow mechanism may occur in the absence of metabolic energy.

Mediated Uptake

All the experimental findings suggest the concept of mediated transport which states that the transport is accelerated due to the presence of carrier substances in the membrane, which interact with the transported ions or molecules. Ions form a complex with the carrier on the outer side of the membrane. This complex is broken down on the inner side.

The following reaction takes place inside the membrane

C + S = CS where S = Substance and C = Carrier.

The concept has been supported by the (1) radioactive ion exchange, (2) saturation effect, and (3) specificity.

(a) Radioactive Ion Exchange:

Using radioactive ions, Legget and Epstein (1956) observed that the ions failed to diffuse freely through the cell membrane. From this observation they pointed out that the movement of ions across the impermeable membrane might be accomplished by the intervention of carriers.

(b) Saturation Effect:

It has been observed that with the increase in salt concentration of the medium, the rates of ion absorption do not increase beyond limit. That is, a saturation point is reached. This is very analogous with the saturation effect found in enzyme catalyzed reactions, suggesting that all active sites of the carriers are occupied by the ions.

At this point, ion transport is kept constant and cannot be made to proceed faster by an increase in salt concentration. So, the phenomenon of saturation effect indicates the presence of carriers.

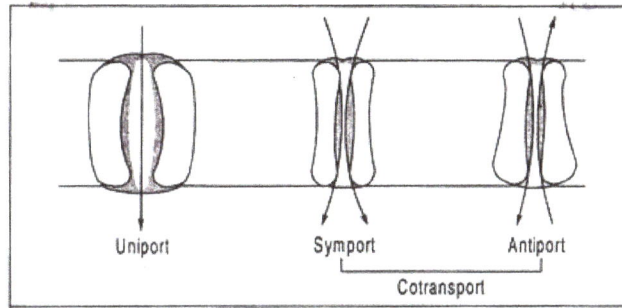

Fig. Uniport, symport and antiport translocation systems found in membranes

(c) Specificity:

Roots absorb ions selectively. Different ions are absorbed at different rates and show different levels of accumulation in the root tissue. This situation is analogous to enzyme-substrate activity. Epstein and Hagen (1952) have observed that the monovalent cations potassium, cesium and rubidium compete with each other for the same binding site.

There are different carriers for different cations and anions. They are also called transporters or perm-eases. They may be of different types such as uniport, symport and antiport. Transporters that carry only one substrate are called uniport systems.

On the other hand, the transporters may carry two ions or solutes simultaneously across a membrane, which are called co-transport systems. When two substrates move in opposite directions, the process is anti-port transport whereas, in symport system, two substrates are moved simultaneously in the same direction.

Mediated transport is classified into two categories depending on the thermodynamics of the system:

1. Passive-mediated transport, or facilitated diffusion, in which a specific molecule flows from high concentration to low concentration.

2. Active transport, in which a specific molecule is transported from low concentration to high concentration, that is, against its concentration gradient. Such an endergonic (energy requiring) process must be coupled to a sufficiently exergonic (energy generating) process to make it favorable (i.e., $\Delta G < 0$).

(d) Passive Mediated Transport:

Ionophores Facilitate Ion Diffusion

One type of carrier molecule is an ionophore, an organic molecule.

Carrier ionophores increase the permeability of membranes to a particular ion by binding the ion, diffusing through the membrane, and releasing it on the other side. For net transport to occur, the uncomplexed ionophore must then return to the original side of the membrane ready to repeat the process (Fig.). The ionic complexes of all carriers must therefore be soluble in nonpolar solvents.

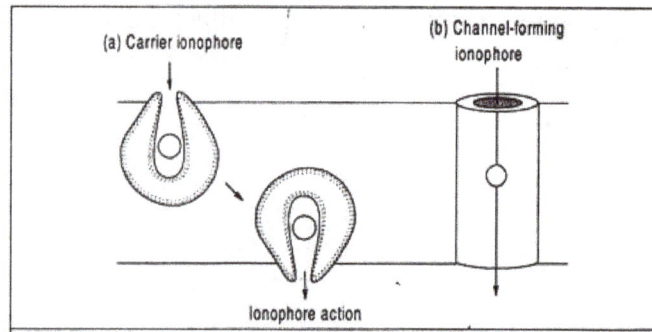

Fig. Carrier ionophores transport ions by diffusing through membrane lipids and channel-forming ionophores span the membrane with a channel through which ions can move

Another type of ionophore called channel-forming ionophores, form solvent-filled, trans membrane channels or pores through which their selected ions can diffuse. Even small amounts of an ionophore greatly increase the permeability of a membrane toward a specific ion.

For example, a single molecule of carrier antibiotic valinomycin transports up to 104K+ ions per second across a membrane. Channel formers, such as the antibiotic gramicidin A, have an even greater ion throughput, over 107K+ ions per second.

Four fundamental classes of transport system are present at all membranes. These are — (1) carriers, (2) pumps, (3) ion channels arid (4) aquaporin's.

Carriers

Unlike pumps, carriers do not catalyse scalar reactions such as ATP hydrolysis.

In other words, the transport process does not involve chemical modification of any of the compounds bound to the carrier. The carriers catalyse only vectorial reactions, i.e., the movement of inorganic ions and simple organic solutes across membranes. The carriers exhibit Michaelis-Menten kinetics that indicates conformational changes during transport.

The array of ions and solutes trans-located by carriers is vast. The principal inorganic nutrients, including NH_4^+, NO_3-, Pi, K^+, and SO_4^{2-}, CI^- are all trans-located into cells by plasma membrane carriers.

The organic solutes like sugars, amino acids, and purine and pyrimidine bases are also trans-located into cells by carriers. The plasma membrane carriers are not only important for nutrient absorption from the soil but also play fundamental roles in the mobilization and storage of metabolites.

For Example

1. The sucrose carriers those are specifically responsible for loading of sucrose into the phloem.

2. Many species reduce NO_3^- in the leaves, and the resulting organic nitrogen compounds are loaded through the carriers into the phloem for transport to sink tissues.

3. During germination, hydrolysis of storage polymers yields sugars and amino acids that need carrier-facilitated mobilization to the growing embryo.

4. Tonoplast carriers catalyse sequestration (temporary storing) of Na^+, Ca^{2+}, Mg^{2+}, and NO_3- as well as sucrose and amino acids into the vacuoles.

5. The triose phosphate trans-locator in the chloroplast membrane exchanges equal amounts of triose phosphate (stromal) for Pi (cytosolic).

6. A carrier in the inner mitochondrial membrane executes the stoichiometric exchange of matrix ATP for cytosolic ADP, maintaining the substrate-product ratio at values favorable for mitochondrial ATP synthesis.

Like the enzymes the carriers are specific for their substrates. Even they are stereo-specific, i.e., specific for D- and L-forms. The acidic, basic and neutral amino acids are transported by distinct plasma membrane carriers. Thus a wide range of carrier types are found in all membranes.

Most of these plant carriers are energized by coupling to the proton motive force (pmf). They behave as H+ – symporters or –anti-porters. Molecular identification of carriers defines them as members of the major facilitator super family. The carriers can be identified genetically.

The most widely applicable and successful approach applied to plant carriers has been yeast complementation. Yeast transport mutants that are defective in growth on the solute of interest are transformed with cDNA from a plant library.

Transformed yeast colonies that are able to grow on the particular solute must accordingly contain functional copies of the relevant plant transport system, which can then be sequenced from the vector. In this way plasma membrane carrier systems for sugars, amino acids, inorganic cations (e.g., K^+), and inorganic anions (e.g., SO_4^{2-}) have been identified in plants.

All the plant carriers identified so far are hydrophobic and contain only single subunit with molecular masses 40 – 50 kDa. The structure of the carrier shows 12 trans membrane spans, with the most extensive hydrophilic loop appearing between trans membrane spans 6 and 7.

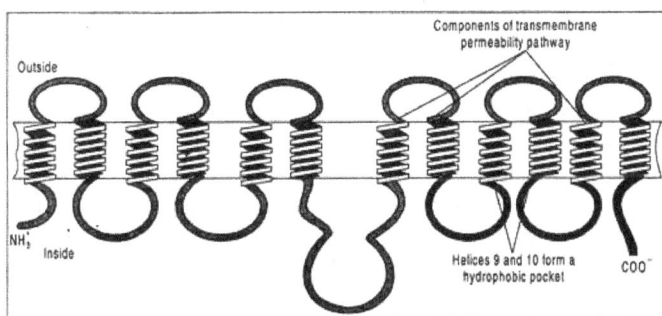

Fig. A generalized structural model of a carrier in a membrane

The motifs present in the proteins are separated by the large hydrophilic loops, indicating that an ancient progenitor of these carrier systems underwent a gene duplication and fusion. Sequence analysis and the general membrane distribution; suggest to place these plant plasma membrane carriers into a large and diverse group of transport systems known as the major facilitator superfamily (MFS).

This group includes both H^+-coupled sugar transport systems from bacteria and uniporters, carriers from animal cells. Expression of such carriers in particular cell type's gives clues to cell function.

Ion Channels

Ion channels are ubiquitous in plant membranes and are studied by electrophysiological techniques. Ion fluxes through channels are driven solely by electrochemical potential differences. Ion flow through channels is passive. Thus, in contrast to pumps or ion-coupled carrier activity, the direction of flow of a particular ion through a channel is dictated by the electrochemical potential gradient for that ion.

The ion channels exhibit ionic substrate selectivity. Most classes of ion channels in plants are of two types — (1) cation channels and (2) anion channels. Cation channels are further subdivided into K^+-selective and Ca^{2+}-selective channels. Most plasma membrane anion channels allow a wide range of anions, including CI^-, NO_3^-, and organic acids. Other anion channels in the tonoplast select specifically for malate.

Ion fluxes through channels are monitored as electrical currents. The selectivity of ion channels is due to the specific binding sites located within the channel pores. In some cases, these ion binding sites have been identified at a molecular level.

Ion channels are gated, often by voltage or ligands, through changes in open state probability. The channels are tightly controlled by conformational shifts between permeable (open, O) and non-permeable (closed, C) states.

This alteration between O and C states, known as gating, is represented by the following reaction:

$$Channel_c \leftrightarrow Channel_Q$$

In all types of channels gating is controlled by membrane voltage, a ligand, or both. When a gating factor activates a channel, the equilibrium of the above reaction shifts from left to right. Voltage-dependent K^+ channels at the plasma membrane allow controlled K^+ uptake and loss. Time-dependent inward and outward currents are carried by separate classes of ion channel.

The channels carrying these currents are called rectifiers. Like valves, rectifying channels carry current in one direction but not the other. For these reasons, the channels are known as K^+ inward rectifiers and K^+ outward rectifiers.

Inward rectifiers take up K^+ ions not only from the soil but also from the apoplast surrounding most cells. The resulting K^+ accumulation contributes to cell turgor. Inward rectifiers have been cloned by complementation of yeast mutants defective in K^+ uptake.

Plant inward rectifier subunits are products of a multi-gene family, members of which show tissue-specific expression. One member, KAT 1, is expressed selectively in guard cells, whereas another AKT1 is expressed in roots and hydathodes.

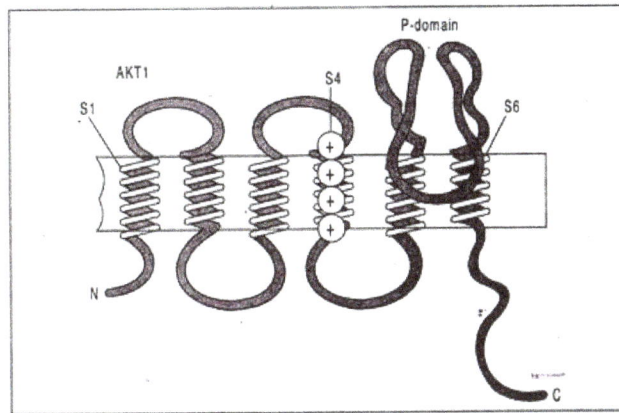

Fig. Structure of inward rectifying K* channel AKT1. S1.S6, there are
six trans membrane apans with a membrane-intrusive loop between S5 and S6,
forming a pore domain (P-domain) of the channel

Several structural features define plant inward rectifying channels as members of the Shaker family, a superfamily of voltage-gated K^+, Ca^{2+}, and Na^+ channels.

The fourth trans membrane helix (S4 domain) of AKT 1 exhibits a regular pattern of positively charged lysine or arginine residues that tend to project from one side of the helix. This region of the protein forms the voltage sensor, which is involved in opening the channel in response to permissive voltage.

Due to hyperpolarization of the membrane, the S4 helix is thought to screw out of the membrane slightly. This conformational transition opens the "gate" that controls ion flow through a separate part of the protein.

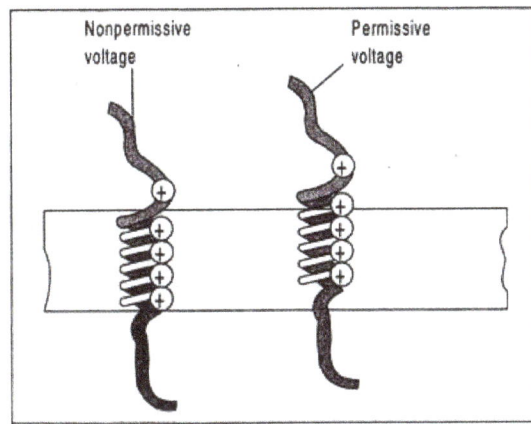

Fig. The voltage sensing S4 domain screws out of the membrane in response to application of an activating voltage

The voltage-insensitive cation channels are major pathways for Na^+ uptake across the plasma membrane and for salt release into the xylem. Such channels give plants tolerance to salinity. These channels are partially blocked by external Ca^{2+}.

Active Uptake of Ion Absorption (Metabolic)

Active uptake of ions is one of the most important features of life processes. It is accomplished through the coupling of diffusion fluxes to the exergonic reactions that take place in the bulk of the membrane. The transfer of ions occurs at the expense of the free energy liberated in chemical reactions. As a rule, this is the energy of hydrolysis of ATP.

It has been observed that both anions and cations are accumulated by plants against concentration gradients to a great extent that cannot be explained by any known electrochemical mechanism, such as ion exchange, Donnan equilibrium, etc.

It has also been observed that the rate and amount of absorption of salts is directly related to the expenditure of metabolic energy. Hober (1945), reported that fresh water alga Nitella, absorbed potassium ions to a concentration 1000 times greater than the concentration of K^+ ions in the surrounding medium.

This type of absorption where the concentration of ions is much higher within the cells than in the external solution is called accumulation.

The extent to which the concentration becomes greater internally than externally is called the accumulation ratio. Thermodynamic laws show that free diffusion and other passive mechanisms not involving expenditure of metabolic energy could not be responsible for such great accumulation.

Many studies indicate that solute transport into cells is strongly dependent upon metabolic energy. Ion accumulation is inhibited when the metabolic activity of the plant is inhibited by low temperatures, low oxygen tension, metabolic inhibitors, and so on. Many workers like Steward (1932), Hopkins (1956) and others have found a parallel curve for salt accumulation and respiration as a function of oxygen tension.

With the decrease in oxygen content of the medium, ion accumulation decreases and ultimately stops completely. Robertson and Turner (1945) and Lundegardh (1955) have reported that metabolic inhibitors like azides and cyanides inhibit ion accumulation with the inhibition of respiration.

An additional evidence in support of the existence of an active process of ion uptake is provided by the phenomenon of salt induced respiration. Lundegardh and Burstrom (1933) observed that the rate of respiration increases when a plant or tissue is transferred from water to a salt solution. The amount by which respiration is increased over the normal rate has been called salt respiration or anion respiration.

Mechanism of Active Uptake

Now, it is necessary to evaluate the concentration effect and the electro potential effect to understand whether the ion absorption takes place under the influence of metabolic energy.

The electrochemical potential difference across the membrane is designated as $\Delta\mu$ and expressed by the following modified Nernst equation:

$\Delta\mu = \Delta (RT \ln C) + \Delta (zFE)$ ['z' is the algebraic valency]

where $\Delta (RT \ln C)$ = chemical potential difference due to concentration effect

$\Delta (zFE)$ = electro potential difference

R = universal gas constant

T = absolute temperature

C = ion concentration inside and outside the membrane

F = Faraday's constant

E = electro potential in volts on each side of the membrane.

The Above Equation Can be Simplified as Follows:

Log Ci/Co = -zFΔE/2.3RT

where Ci = inside ion concentration

Co = outside ion concentration

Ci/Co is the predicated ratio at equilibrium, when μ = 0. If the actual-ratio is greater than the predicted ratio, the cell is performing work to absorb ions using energy. This is called active absorption. Passive absorption occurs when the actual ratio is equal to or lesser than the predicated value.

The active ion transport may be (1) primary and (2) secondary.

Primary Active Transport

The primary active transport is coupled directly to a source of energy other than electrochemical potential gradient, such as ATP hydrolysis, an oxidation reduction reaction, etc. The membrane proteins that carry out primary active transport are called pumps. Most pumps transport ions. Ion pumps are further characterized as either electro genic or electro neutral.

In general, electro genic transport refers to ion transport involving the net movement of charge across the membrane. In contrast, electro neutral transport, as the name implies, involve no net movement of charge. For example, the Na^+/K^+ -ATPase of animal cells pumps three Na^+ ions out of every two K^+ ions in, resulting in a net outward movement of one positive charge.

Secondary Active Transport

Secondary active transport uses the energy stored in electrochemical-potential gradient. Protons are extruded from the cytosol by electro genic H^+ ATPase operating in the plasma membrane and tonoplast. Consequently, a membrane potential and a pH gradient are created at the expense of ATP hydrolysis.

This gradient of electrochemical potential for H^+, the proton motive force represents stored free energy in the form of H^+ gradient.

Carbon Cycle in Soil

Carbon (C) is one of the most common elements in the universe and found virtually everywhere on earth: in the air, the oceans, soil, and rock. Carbon is part of geologic history in rock and especially the ancient deposits that formed coal, oil and other energy sources we use today. Carbon is also an essential building block of life and a component of all plants and animals on the planet. It has unique bonding properties that allow it to combine with many other elements. These properties enable the formation of molecules that are useful and necessary to support life. The role of carbon in living systems is so significant that a whole branch of study is devoted to it: organic chemistry. Carbon that is not tied up in rock or deep in the oceans is constantly changing and moving. This process is called the carbon cycle. Soil holds the largest portion of active carbon on earth. Plants take carbon from the air and convert it to plant tissue, some of which returns to the soil as plant residue.

Agriculture's Role in the Carbon Cycle

Carbon is critical to soil function and productivity, and a main component of and contributor to healthy soil conditions. Soil management plays a critical role in whether the carbon remains in the soil or is released to the atmosphere. Agricultural practices can impact both the amount and the composition of soil organic carbon and hence also the soil's physical, biological, and chemical condition, the combination of things that defines soil health. Farm practices that affect carbon therefore impact agricultural productivity and resilience (the soil's ability to deal with weather extremes) and the carbon cycle itself.

Importance of Soil Organic Carbon

While the agricultural sector has the ability to impact the carbon cycle on a large scale, often through the release of carbon, farmers have a vested interest in retaining and increasing soil organic carbon for individual fields because soil and yield tend to improve when the soil organic carbon level increases. Higher soil organic carbon promotes soil structure or tilth meaning there is greater physical stability. This improves soil aeration (oxygen in the soil) and water drainage and retention, and reduces the risk of erosion and nutrient leaching. Soil organic carbon is also important to chemical composition and biological productivity, including fertility and nutrient holding capacity of a field. As carbon stores in the soil increase, carbon is "sequestered", and the risk of loss of other nutrients through erosion and leaching is reduced. An increase in soil organic carbon typically results in a more stable carbon cycle and enhanced overall agricultural productivity, while physical disturbances of the soil can lead to a net loss of carbon into the surrounding environment due to formation of carbon dioxide (CO_2).

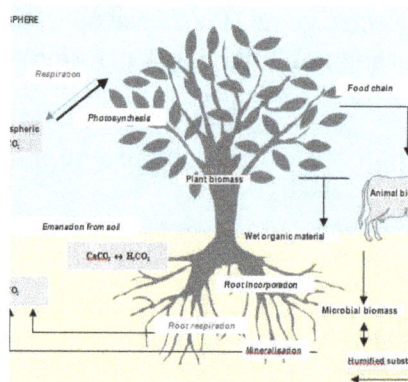

Figure. The carbon cycle and the interaction of plant, soil, and atmosphere.

Management Practices for C Sequestration

With agricultural productivity so dependent on soil organic carbon and carbon cycling, how can we best manage fields to enhance soil organic carbon levels while also reducing carbon loss into the atmosphere?

Table. Management practices that can increase soil organic carbon and reduce carbon loss into the atmosphere.

Management practices	Functions and explanation
Conservation tillage practices	Conservation tillage practices including no-till management aid in storing soil organic carbon, keeping the physical stability of the soil intact. When reduced-till systems are combined with residue management and manure management, soil organic carbon can increase over time.
Crop residue management	Returning crop residue to the soil adds carbon and helps to maintain soil organic matter.
Cover crops	Cover crops can increase soil carbon pools by adding both root and above ground biomass. Covers also reduce the risk of soil erosion and the resulting loss of carbon with soil particles. Cover crops also enhance nutrient cycling and increase soil health over time.
Manure and compost	Adding organic amendments such as manure or compost can directly increase soil carbon, and also result in increased soil aggregate stability. This enhances the biological buffering capacity of the soil, resulting in greater yields and yield stability over time.
Crop selection	Perennial crops eliminate the need for yearly planting and increase soil organic carbon by root and litter decomposition post-harvest. Crops with greater root mass in general add to root decomposition and physically bond aggregates together. Using high residue annual crops can also help reduce net carbon loss from cropping systems.

The ability of agricultural fields to sequester carbon (capture and storage of carbon that would otherwise be lost to the environment) depends on several factors including climate, soil type, type of crop or vegetation cover, and management practices. Employing farming practices that reduce disturbance of the soil (less aeration from tillage helps protect carbon), combined with practices that bring additional carbon to the soil, will allow for carbon sequestration over time. Such practices include implementation of conservation tillage (no-till, zone-till, minimum-till, shallow mixing or injection for manure applications), retaining crop residues, including cover crops in crop rotations, adding organic nutrient sources such as manure and compost, and including perennial crops in crop rotations (Table). Their implementation may slow or even reverse the loss of carbon from agricultural fields, improve nutrient cycling and reduce nutrient loss.

Soil carbon management is an important strategy for improving soil quality, increasing crop yields, and reducing soil loss. Capturing carbon in the soil helps improve soil health and productivity, and stabilize the global carbon cycle, benefiting agricultural production.

Nitrogen Cycle in Soil

Nitrogen exists in soils in many forms and constantly changes from one form to another. The paths that the different forms of nitrogen follow through the ecosystem are collectively called the nitrogen cycle. Understanding how the different pools of nitrogen interact and the processes by which these forms enter and leave the cycle is the subject of continuing study. Nitrogen is found in both inorganic and organic forms. Soil organic matter is composed primarily of amides (NH_2) and accounts for more than 90 percent of the total nitrogen present in most environments. In general, nitrogen is not found associated with soil minerals, as is the case with phosphorus. Some clay minerals may tie up small amounts of nitrogen in the ammonium form (clay fixation) but not in the same magnitude as phosphorus.

Organic Nitrogen

Soil organic matter is the major storehouse of many plant nutrients in soils, including nitrogen, phosphorus, sulfur, calcium and magnesium. Soil organic matter is composed of a stable material called humus, an easily decomposed material (litter), soil microbes and some other organic molecules. Typically, humus will contain 45 to 55 percent carbon and about 5 percent nitrogen. In other words, soil humus typically has a carbon to nitrogen ratio (C:N ratio) of approximately 12:1 for surface soils and 8-10:1 for subsurface soils.

Mineralization

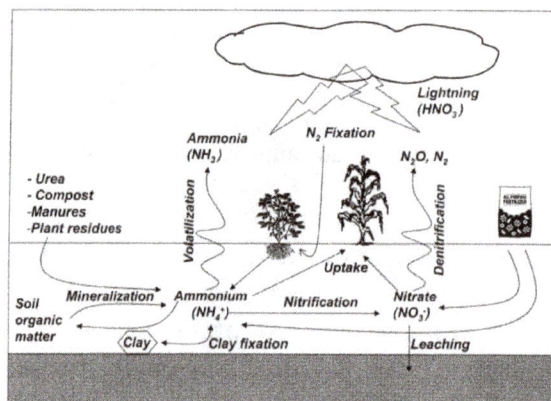

Figure. Nitrogen cycle in upland soils

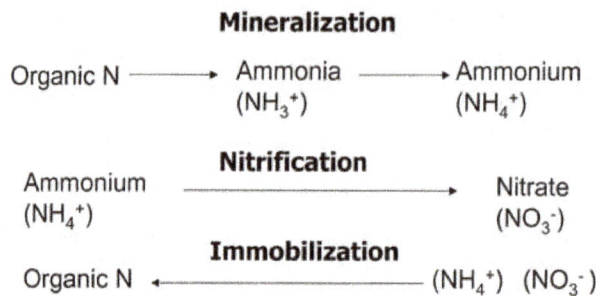

Figure. Nitrogen mineralization/immobilization process

Nitrogen that is present in soil organic matter, crop residues and manures is converted to the inorganic form by the process of mineralization. Initially, larger organic matter molecules are broken down into smaller ones, with soil microorganisms attacking these remaining materials by producing specific enzymes. The transformation of organic nitrogen to the ammonia (NH_3) and ammonium (NH_4^+) forms is referred to as ammonification. The resultant ammonium can then be transformed to the nitrate form (NO_3^-) by a process called nitrification. Since the decomposition process is carried out by living organisms, it is affected by several environmental variables, including soil moisture, temperature, pH, the C:N ratio and the type of organic materials in the residue. Cotton and grain sorghum stalks and small grain residues are relatively high in carbon and low in nitrogen (having C:N ratios greater than 30:1). Soil organisms may need additional nitrogen to decompose these residues. When the nitrogen supply is limited, soil microbes compete with plants for fertilizer N in a process called immobilization. When these microbes die, nitrogen tied up in the decomposition process again becomes available for crop use.

Biological Nitrogen Fixation

Atmospheric nitrogen (N_2) is basically an endless source of N, but this nitrogen cannot be used directly by most plants. Legumes such as alfalfa, soybeans and clovers form a symbiotic association (mutually beneficial) with specific bacteria to convert atmospheric N_2 to a form plants can use. The plant provides nutrients and other compounds to the N_2-fixing bacteria, and in return, the plant benefits from the N fixed by the microorganisms. The amount of N fixed varies among plants and growing conditions (Table). The symbiotic association is highly specific; thus, the bacterial species that fixes N2 with soybeans is not effective for fixing N_2 with alfalfa. The site of the N_2-fixing process is a root nodule that forms on the root system and has a pink color if actively fixing N_2.

The biological nitrogen fixation process is catalyzed (accelerated) by the nitrogenase enzyme, which is affected by a number of soil and weather factors. The association does not work well in soils with very low pH or when high levels of available mineral nitrogen are present. The absence of the right type of inoculum (bacteria) in the soil also will limit nodulation and N_2 fixation.

Table: Typical Amount of Nitrogen Fixed by Selected Legumes

Crop	Amount fixed (lb/A/year)
Alfalfa	170-270
Soybeans	70-170
Clovers	120-170
Vetch	70-150

Some other soil organisms are capable of fixing N2 via non-symbiotic associations. This process is of little significance in most agricultural systems, with the exception of blue-green algae which live in rice systems. Some nitrogen also is fixed by lightning, but in considerably smaller amounts than biologically fixed N_2.

Leaching

Nitrogen in the nitrate form is very mobile and highly soluble in water. Rainfall moving through the root zone may wash nitrate downward, reaching tiles or drainage channels and potentially reaching groundwater or surface waters. Leaching is a more serious problem in highly permeable sandy soils than in clayey soils. The magnitude of nitrate loss through leaching depends on the amount and intensity of the rainfall or irrigation water and the amount of nitrate present in the soil.

Loss of nitrate by leaching is of concern for two important reasons. Nitrate below the root zone is no longer available for plant use, representing an important loss of resources. Water quality problems caused by excess nitrogen leaving fields may result in the deterioration of drinking water sources and wildlife habitat.

Volatilization as NH3

A significant amount of nitrogen fertilizer can be lost through the process of volatilization if not properly applied. Many commercial fertilizers contain ammonium or convert readily to ammonium, such as urea. Under alkaline pH conditions (pH > 7), a percentage of the ammonium (NH_4^+) can be converted to ammonia gas (NH_3) and escape to the atmosphere. In addition to soil pH, high soil temperature, excessive soil moisture and strong winds will contribute to such loss.

Ammonia volatilization also can occur under neutral and acidic soil pH conditions. During the transformation of urea to ammonium (NH_4^+), the pH of soil adjacent to the fertilizer can increase to levels that promote ammonia formation. This effect is more pronounced in sandy and silt loam soils, which lack the ability to resist pH changes (i.e., low pH buffering capacity).

Denitrification

The process of denitrification is a form of respiration carried out by microorganisms under low oxygen conditions. Specific microorganisms, primarily bacteria, have the ability to use nitrate instead of oxygen to carry out their metabolic functions. In this process, nitrate (NO_3^-) is reduced to NO_2^- and then to gaseous forms including N_2O and N_2.

Temperatures between 75 and 95 degrees F, low soil oxygen levels for several days (waterlogged conditions), soil pH > 6 and the presence of a carbon source (organic matter) are ideal conditions for denitrification to proceed at a high rate. Denitrification varies widely among locations and time of year, but it normally represents only a small percentage of total N loss. The process of denitrification is detrimental, since plant-available nitrogen is lost to the atmosphere. However, denitrification also can be used as a means to prevent the buildup of nitrates in groundwater wells, manure storage lagoons and wastewater treatment plants.

Nitrogen Cycle in Flooded Soils

In upland soils, the process of mineralization gradually converts organic nitrogen to ammonium and then to nitrate. Under flooded conditions, due to the low oxygen concentration, the end product of the mineralization process is ammonium (Figure).

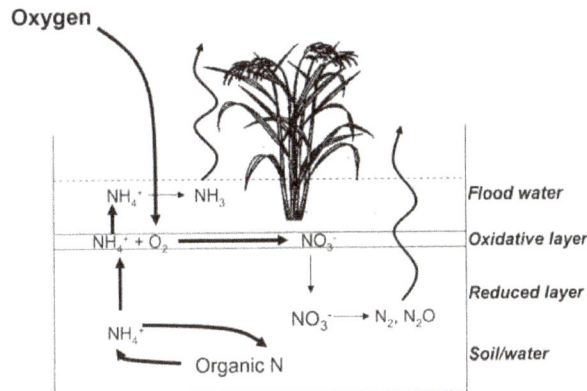

Figure. Simplified nitrogen cycle in flooded soils

In flooded soils, such as rice paddies, two layers are formed – an oxidative layer at the soil/water interface, which is less than one inch thick, and a reductive layer beginning immediately beneath. If nitrate fertilizer is added to flooded soil, bacteria can convert it to nitrogen gas (denitrification), which may be lost to the atmosphere. Ammonium fertilizer placed in the oxidative layer also may be converted to nitrates, just as in an upland soil.

Phosphorous Cycle in Soil

Phosphorus, Crops and the Environment

This agronomy fact sheet provides a brief overview of the important components of the phosphorus (P) cycle. Understanding the P cycle can help producers make decisions regarding P management on the farm, both for farm profitability and protection of the environment.

Most plants are only about 0.2% P by weight, but that small amount is critically important. Phosphorus is an essential component of adenosine triphosphate (ATP), which is involved in most biochemical processes in plants and enables them to extract nutrients from the soil. Phosphorus also plays a critical role in cell development and DNA formation. Insufficient soil P can result in delayed crop maturity, reduced flower development, low seed quality, and decreased crop yield. Too much P, on the other hand, can be harmful in some situations; when P levels increase in fresh water streams and lakes, algae blooms can occur. When algae die, their decomposition results in oxygen depletion which can lead to the death of aquatic plants and animals. This process is called "eutrophication".

Crop Uptake

One goal with field crop management is to optimize crop uptake of available P. A typical corn silage crop will remove about 4.3 lbs of P_2O_5 per ton of silage (35% dry matter). Soil testing of available P can help avoid application of fertilizer P that is not needed for optimum production. Applying fertilizer beyond crop needs is a waste of time and money, and can be harmful to the environment.

Phosphorus Cycle

Phosphorus exists in many different forms in soil. For practical purposes, we can group these sources into four general forms: (1) plant available inorganic P, and three forms which are not plant available: (2) organic P, (3) adsorbed P, and (4) primary mineral P. The P cycle in Figure shows these P forms and the pathways by which P may be taken up by plants or leave the site as P runoff or leaching. The general P transformation processes are: weathering and precipitation, mineralization and immobilization, and adsorption and desorption. Weathering, mineralization and desorption increase plant available P. Immobilization, precipitation and adsorption decrease plant available P.

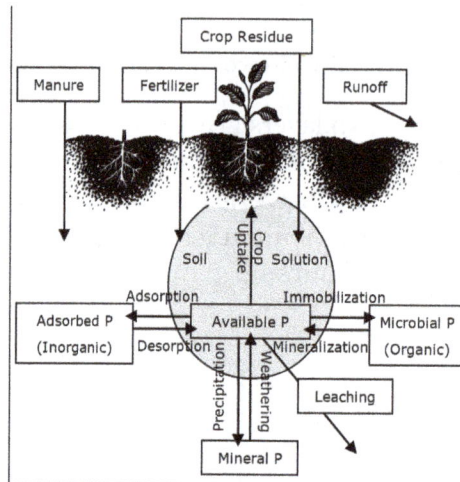

Figure: Simplified phosphorus cycle.

Weathering and Precipitation

Soils naturally contain P-rich minerals, which are weathered over long periods of time and slowly made available to plants. Phosphorus can become unavailable through precipitation, which happens if plant available inorganic P reacts with dissolved iron, aluminum, manganese (in acid soils), or calcium (in alkaline soils) to form phosphate minerals.

Mineralization and Immobilization

Mineralization is the microbial conversion of organic P to $H_2PO_4^-$ or HPO_4^{2-}, forms of plant available P known as orthophosphates.

Immobilization occurs when these plant available P forms are consumed by microbes, turning the P into organic P forms that are not available to plants. The microbial P will become available over time as the microbes die.

- Maintaining soil organic matter levels is important in P management. Mineralization of organic matter results in the slow release of P to the soil solution during the growing season, making it available for plant uptake. This process reduces the need for fertilizer applications and the risk of runoff and leaching that may result from additional P.

- Soil temperatures between 65 and 105°F favor P mineralization.

Adsorption and Desorption

Adsorption is the chemical binding of plant available P to soil particles, which makes it unavailable to plants. Desorption is the release of adsorbed P from its bound state into the soil solution.

- Adsorption (or "fixing" as it is sometimes called) occurs quickly whereas desorption is usually a slow process.

- Adsorption differs from precipitation: adsorption is reversible chemical binding of P to soil particles while precipitation involves a more permanent change in the chemical properties of the P as it is removed from the soil solution.

- Soils that have higher iron and/or aluminum contents have the potential to adsorb more P than other soils.

- Phosphorus is in its most plant available form when the pH is between 6 and 7. At higher pH, P can precipitate with Ca. At lower pH, P tends to be sorbed to Fe and Al compounds in the soil.

- Every soil has a maximum amount of P that it can adsorb. Phosphorus losses to the environment through runoff and/or leaching increase with P saturation level.

- Precise fertilizer placement can decrease P adsorption effects by minimizing P contact with soil and concentrating P into a smaller area. Band application of fertilizer is a common example of this.

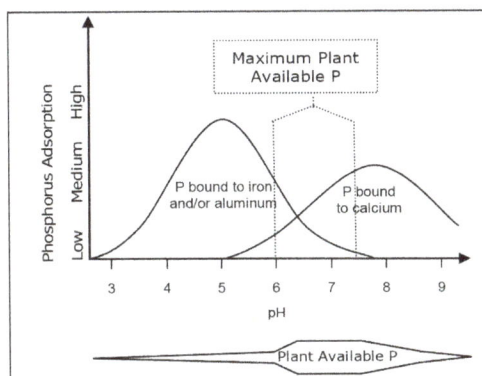

Figure: Soil pH impacts P availability.

Runoff

Runoff is a major cause of P loss from farms. Water carries away particulate (soil-bound) P in eroded sediment, as well as dissolved P from applied manure and fertilizers. Erosion control practices decrease P losses by slowing water flow over the soil surface and increasing infiltration.

Leaching

Leaching is the removal of dissolved P from soil by vertical water movement. Leaching is a concern in relatively high P soils (near or at P saturation), especially where preferential flow or direct connections with tile drains exist.

Crop uptake is the goal of applying P fertilizer or manure to the soil. If soil tests P levels are already optimum, P additions through fertilizer or manure should not exceed crop removal. If additional P is needed (soils testing low or medium in P), P adsorption can be minimized by band applications and by maintaining an optimum pH. Naturally occurring immobilization of P by microbes can help ration plant available P to crops over the course of a growing season. Steps should be taken to reduce losses in order to maximize the efficiency of fertilizer and manure applications.

Potassium Cycle in Soil

Potassium: Nature and Ecological Roles: Potassium is an essential plant nutrient that sometimes limits plant growth. Potassium is only present in soil as a positively charged cation (K+). Its entire lifecycle in the soil is actually linked mostly with cation exchange and mineral weathering. Potassium doesn't cause any nutrient pollution problems and is generally pretty inert.

Potassium in Plant and Animal Nutrition: Potassiums main function in plants and animals is actually not to be synthesized into organic compounds, rather, it activates enzymes. It is the activator for over 80 enzymes responsible for living functions. Potassium always remains in the ionic form as well. Potassium is linked to plants having a really good response to environmental stress as well. It improves a plants hardiness, drought tolerance, and disease resistance. This is an essential nutrient to help plants flower or produce fruit. Potassium is necessary for animals including humans and helps regulate the central nervous system and maintain healthy blood vessels.

Deficiency Symptoms in Plants: Potassium deficiency in plants creates a foliar yellowing on the tips and edges of the oldest leaves first due to its mobility in plants. Certain plants, mainly legumes, will get while spots towards the tips of their leaves. To tell it apart from salinity damage looks for these symptoms in old vs new growth.

The Potassium Cycle: Potassium comes from primary minerals in soil like micas and potassium feldspar. These minerals slowly weather and eventually the potassium becomes more and more available to be held in soil solution or taken up by plant roots. Once plants take up a ton of potassium it is leached from leaves to soil by rainfall. This and animal urine is how it gets returned to the soil. Some potassium is lost to soil losses, runoff, and leaching to groundwater. When plants are harvested and taken away from the soil entirely, the nutrients including potassium goes to wherever the plant matter goes thus ag. soils need potassium amendments. Soils amendments high in potassium include poultry manure and wood ash.

The Potassium Problem in Soil Fertility

Availability of Potassium: Most mineral soils hold a larger volume of potassium than phosphorus and actually the total quantity of soil potassium is larger than any other major nutrient. However, available potassium is found in small amounts in the soil.

Leaching Losses: Potassium is mostly lost from the soil by leaching however its positive structure is great for binding to negative cation exchange sites on clay or humus. Liming can help keep potassium in the soil due to complementary ion effect (K^+ ions can more easily replace Ca^{2+} ions in cation exchange).

Plant Uptake and Removal: Plants take up about as much potassium as they do nitrogen. The removal mostly occurs when plants are harvested and most of the plant matter is removed from the soil it was grown.

Luxury Consumption: Plants will actually take up more potassium than they need when available quantities are large. This means that anything a plant takes up above its required potassium will be wasted if the plant is removed from where it grew. High potassium levels in plants could actually inhibit calcium and magnesium uptake causing nutritional imbalance.

Forms and Availability of Potassium in Soils: There are four forms of potassium involved in the soil potassium cycle: 1. K in mineral structures (unavailable), 2. K in nonexchangeable mineral forms (slowly available), 3. Exchangeable K on colloid surfaces (readily available), and 4. Water soluble K ions (readily available). The exchange of potassium between these four forms is a function of the types of clay minerals. Soils with 2:1 clays have the most potassium. Some plants can actually obtain potassium from generally unavailable forms on primary mineral structures.

Relatively Unavailable Forms: This includes about 90-98% of all soil potassium. This is the potassium held up in feldspar or mica. This form can release potassium slowly over the course of many years. This weathering of the primary potassium minerals is generally assisted by organic/inorganic acids and acidic clays and humus.

Readily Available Forms: Only about 1-2% of total soil potassium. This form of K exists in two forms: 1. In the soil solution, 2. Exchangeable on colloid surfaces. Most of this is in the exchangeable form on charged colloid sites. The free K in soil solution is mostly used by higher plants. There is a continued equilibrium in soils which keeps potassium evenly in solution and on exchange sites.

Slowly Available Forms: Sometimes in type 2:1 mineral clay soils, potassium as well as the similarly sized ammonium molecule will get permanently affixed between layers of growing soil colloids. These aren't held on exchange sites and so are "nonexchangeable ions". This however, acts as an important reservoir for slowly released ions.

Release of Fixed Potassium: There is a lot of nonexchangeable/fixed potassium in soils. It is continually released to an exchangeable form in amounts large enough to be important. There is a good equilibrium between exchangeable K and nonexchangeable. This means some soils, especially sandy soils, have lower CEC and thus don't maintain potassium well. More clayey soils have a better time maintaining enough potassium ions throughout a growing season.

Factors Affecting Potassium Fixation in Soils: There are four soil conditions that influence the amount of K that can be fixed: 1. Types of soil colloids, 2. Wet and dry cycles, 3. Freeze and thaw cycles, 4. The presence of lime.

Effects of Type of Clay and Moisture: Type 1:1 clays fix a small amount of potassium while 2:1 clays fix K in large quantities. Freeze/thaw and wet/dry cycles help to stimulate the exchange equilibrium of K.

Influence of pH: Lime application generally increases soil pH and potassium fixation. In acidic soils, colloids are holding H+ ions preventing K+ from getting close to the exchange sites thus keeping K+ from fixating. So as the pH increases, more K is able to be fixed to soil colloids. Another factor is root uptake, the more calcium and magnesium in soils the more competition these cations have for root uptake and thus less potassium could be absorbed.

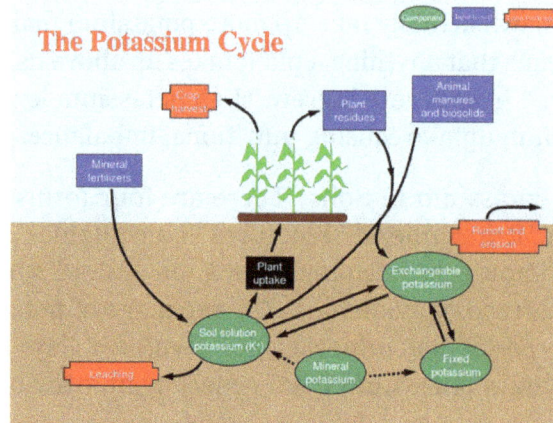

Potassium Cycle

Practical Aspects of Potassium Management: In most soils, the potassium fertility issue comes with the rate at which potassium can be converted to a plant available form. When plant material isn't removed, cycling between plant matter and soil is usually adequate to supply the next season's plants. A tricky thing about potassium management is that K uptake by plants is not consistent throughout the growing season. Many farmers rely on potassium additions with fertilizer however, excessive K depress Ca and Mg which can cause disruption in plant and animal health.

Frequency of Application: The author suggests that it's best to apply small amounts of potassium to fields every so often in order to prevent excess leaching or over absorption of K by plants. Application would generally increase over many years until only maintenance levels were needed, soils would become built up enough to supply a good amount of potassium by themselves.

Chapter 7

Soil Quality

The measure of the soil condition relative to its requirement, whether to sustain plant or animal life, support human health or to enhance air and water quality is termed as soil quality. This chapter has been written in order to help develop a better understanding of soil quality. It includes the topics related to this area, such as soil tests for determining soil quality, soil respiration and bulk density of the soil.

Soil quality is how well soil does what we want it to do. A healthy, high-quality soil is fertile, has good soil structure, and is biologically active.

Soil health is the foundation of productive farming practices. Fertile soil provides essential nutrients to plants. Important physical characteristics of soil structure and aggregation allow water and air to infiltrate, and roots to explore. Soil health and soil quality are terms used interchangeably to describe soils that are not only fertile but also possess beneficial physical and biological properties. What is soil quality? Soil quality is how well soil does what we want it to do. Soil quality is the capacity of a specific kind of soil to function to sustain plant and animal productivity, maintain or enhance water and air quality, and support human health and habitation.

Soil Fertility

Soil fertility is the ability of a soil to provide the nutrients needed by crop plants to grow. The primary nutrients plants take up from soils include nitrogen, phosphorus, potassium, calcium and magnesium. Frequently, we need to supplement soil nutrients by adding fertilizer, manure or compost, for good crop growth. Plants take up many other nutrients from soils, but there is usually enough of these secondary nutrients in the soil, so there is no need to add more.

Soil pH is another important aspect of soil fertility. pH is not a plant nutrient, but rather is a measure of the acidity of the soil. Most crops grow best when the soil pH falls between 6.2 and 6.8. This is the range in which plant roots can best absorb most nutrients from the soil.

Organic Matter

Organic matter is composed of plant and animal residues, living and dead soil microorganisms, and substances produced through decomposition. Most agricultural soils contain only a small proportion of organic matter (usually less than 5%), but this small amount plays a very large role in soil quality. Soil organic matter tends to improve soil fertility, soil structure, and soil biological activity. Organic matter is added to soils through cover crops, manure, compost, and crop rotation.

Soil Texture

Soil texture is an important soil characteristic that influences many aspects of soil quality. The textural class of a soil is determined by the percentage of sand, silt, and clay. Soils are usually made up of a mix of the three particle sizes. Sand particles are relatively large, clay particles are very tiny in comparison to sand, and silt particles are medium-sized. Clay and silt particles hold more water and plant nutrients along their surfaces than sand particles. Soil texture is an inherent property of a soil, and does not change under different management practices. Soils can be classified as one of four major textural classes: (1) sands; (2) silts; (3) loams; and (4) clays. These are based on the proportion of particle sizes found in each soil.

Relative Soil Particle Size

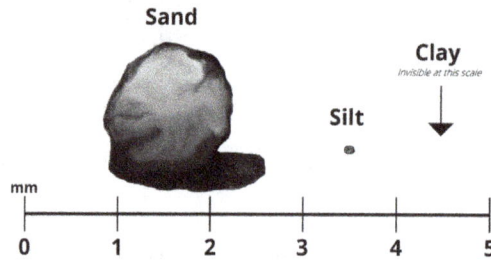

Soil Particles Vary Greatly in Size, As Shown in This Illustration.

Knowing the texture of a soil provides you with quite a bit of information about how well the soil holds water, holds and releases nutrients, and responds to different tillage practices. For example, a clay soil will hold more nutrients and more water compared to a sandy soil, but will be more susceptible to compaction from plowing and cultivating.

While soil texture is the proportion of the three soil particle types (sand, silt, and clay), soil structure refers to how those particles are arranged in space. We cannot change soil texture, but we can manage soils to improve soil structure. Soil with good structure has approximately 40-60% of its volume in pore space, or empty space between soil particles. Water and air can get into these pore spaces, and roots can grow into these spaces.

In a healthy soil, particles of sand, silt and clay aren't floating around by themselves. They are joined up with other particles, bits of organic matter, and small pore spaces into soil aggregates. Stronger, more stable aggregates stick together, even when hit by a raindrop or crushed by a footstep. A handful of healthy soil feels crumbly and light, due mostly to these stable aggregates.

Compaction

Plant growth is limited in compacted soils

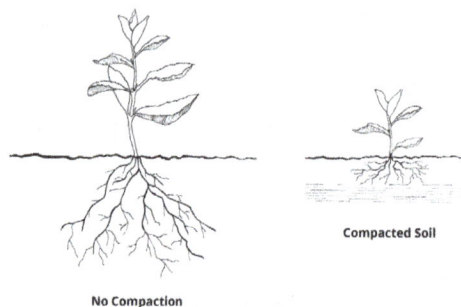

Compacted Soil

No Compaction

Plants do not grow well in compacted soils because there is less space between soil particles for roots to grow into.

Soil compaction occurs when soil aggregates are pushed closer together, and pore spaces shrink. This usually occurs when heavy tractors, trucks and other machines are driven over soil, particularly if soils are wet. Soils can become compacted at the surface, but also at the layer of soil just below the depth of tillage (subsoil compaction). Plants have difficulty growing in compacted soil because the soil aggregates are pressed together, leaving little pore space for air and water, which are essential for root growth.

Water Holding Capacity

Soil water holding capacity is the amount of water that a given soil can hold and then make available for crop use. Water holding capacity is largely determined by soil texture and by the amount of pore spaces in the soil, where water and air can be found. Sandy soils have lower water holding capacity, while silt and clay soils tend to have higher water holding capacity. A crop grown in a sandy soil will need to be irrigated more frequently, but with less total water, than a crop grown in a clay or silty soil. A clay or silty soil will hold more water for the crop to use, so can be irrigated less frequently. Compacted soils have less pore space for the water, and therefore have lower water holding capacity.

Soil Biological Activity

Healthy soils are teeming with living organisms: bacteria, fungi, insects, earthworms, etc. As these living things go through their life cycles, they perform many functions that help improve the quality of soil. Soil organisms decompose fresh organic matter such as crop residues and animal manures. In the process, they help soil particles stick together into stable aggregates. They also create humus, a form of organic matter that doesn't decompose further, that helps soils hold water and nutrients. Soils with higher biological activity tend to have fewer plant disease organisms. Earthworms tunnel through soils, opening up pathways for air and water to move into the soil.

Soil Conservation

When water from rainfall or irrigation washes over bare soil, or wind blows over bare soil, soil particles may be washed or blown away, out of the field. This process is called soil erosion; the farming practices we use to stop erosion are known as soil conservation practices. Healthy soil is a very valuable natural resource, and we don't want to lose soil out of our fields. Soil particles that erode from fields can cause environmental problems, such as polluting creeks, rivers, lakes and even oceans. Airborne soil particles can lower air quality, and cause respiratory illnesses. Farmers can protect soils from erosion by limiting the time when there is bare soil in the field, improving soil structure, and by managing tillage, irrigation and crop rotation.

Soil Test

Atterberg Limits

Consistency is the term used to describe the ability of the soil to resist rupture and deformation. It is commonly describe as soft, stiff or firm, and hard.

Water content greatly affects the engineering behavior of fine-grained soils. In the order of increasing moisture content, a dry soil will exist into four distinct states: from solid state, to semi-solid state, to plastic state, and to liquid state. The water contents at the boundary of these states are known as Atterberg limits. Between the solid and semisolid states is shrinkage limit, between semisolid and plastic states is plastic limit, and between plastic and liquid states is liquid limit.

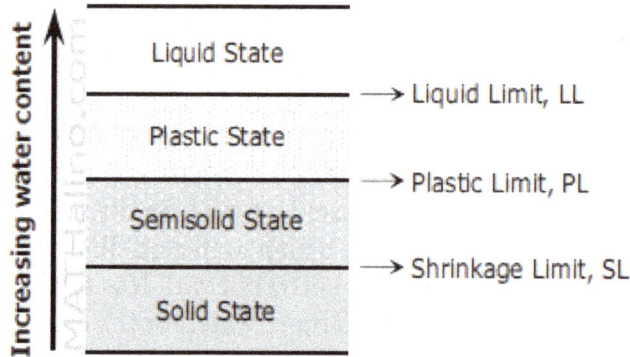

Figure. Atterberg Limits

Atterberg limits, then, are water contents at critical stages of soil behavior. They, together with natural water content, are essential descriptions of fine-grained soils.

Liquid Limit, LL

Liquid limit is the water content of soil in which soil grains are separated by water just enough for the soil mass to loss shear strength. A little higher than this water content will tend the soil to flow like viscous fluid while a little lower will cause the soil to behave as plastic.

Plastic Limit, PL

Plastic limit is the water content in which the soil will pass from plastic state to semi-solid state. Soil can no longer behave as plastic; any change in shape will cause the soil to show visible cracks.

Shrinkage Limit, SL

Shrinkage limit is the water content in which the soil no longer changes in volume regardless of further drying. It is the lowest water content possible for the soil to be completely saturated. Any lower than the shrinkage limit will cause the water to be partially saturated. This is the point in which soil will pass from semi-solid to solid state.

Determination of Liquid, Plastic, and Shrinkage Limits

Casagrande Cup Method for Liquid Limit Test

The semispherical brass cup is repeatedly dropped into a hard rubber base from a height of 10 mm by a cam-operated crank.

The dry powder of the soil is mixed with distilled water turning it into a paste. The soil paste is then placed into the cup to a thickness of about 12.5 mm and a groove is then cut at the center of the paste using the standard grooving tool. The crank operating the cam is turned at the rate of 2 revolutions per second lifting the cup and dropped it from a height of 10 mm. The liquid limit is the moisture content required to close a distance of 12.5 mm along the bottom of the groove after 25 blows.

Casagrande Cup

The required closure in 25 blows is difficult to achieve in a single test. Four or more tests to the same soil at varying water contents are to be done for 12.5 mm closure of the groove. The results are then plotted on a semi-logarithmic graph with moisture content along the vertical axis (algebraic scale) and number of blows along the horizontal axis (logarithmic scale).

The graph is approximated by the best fit straight line, usually called the flow line and sometimes called liquid state line. The moisture content that corresponds to 25 blows is the liquid limit of the soil.

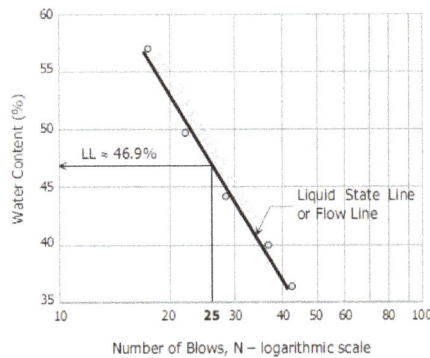

Figure 3 Typical liquid limit results from the Casagrande cup method.

Figure. Typical liquid limit results from the casagrande cup method.

The slope of the flow line is called flow index and may be written as

Flow index, $FI = \dfrac{w_1 - w_2}{\log(N_2/N_1)}$

where w_1 and w_2 are the water content corresponding to number of blows N_1 and N_2, respectively.

Plastic Limit Test

The plastic limit can easily be found by rolling a small soil sample into thin threads until it crumbles. The water content at which the threads break at approximately 3 mm in diameter is the plastic limit. Two or more tests are made and the average water content is taken as plastic limit. In this test, soil will break at smaller diameter when wet and breaks in larger diameter when dry.

Fall Cone Method for Liquid and Plastic Limit Tests

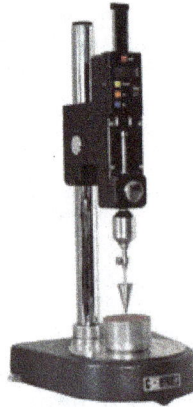

Cone Penetrometer

Fall cone method offers more accurate result of liquid limit and plastic limit tests. In this method, a cone with a mass of 80 grams and an apex angle of 30° is suspended above so that its pointed part will just in contact with the soil sample. The cone is permitted to fall freely under its own weight for a period of 5 seconds. The water content that allows the cone to penetrate for 20 mm during this period defines the liquid limit of the soil.

Like the cup method, four or more tests are required because it is difficult to find the liquid limit in a single test. The results are then plotted into a semi-logarithmic paper with water content along the vertical axis (arithmetic scale) and penetration along the horizontal axis (logarithmic scale). The best fit straight line is then drawn and the water content that corresponds to 20 mm penetration defines the liquid limit.

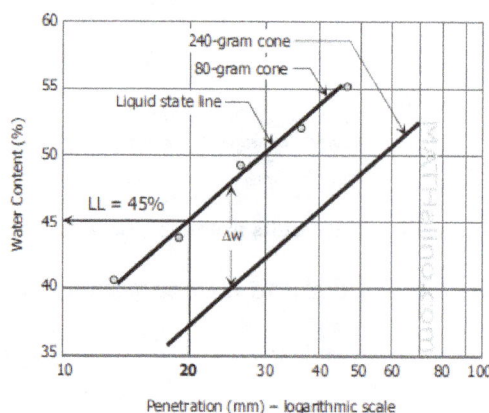

Figure. Typical test results from the fall cone apparatus.

The plastic limit can be found by repeating the test with a cone of similar geometry but with a mass of M2 = 240 grams. The liquid state line of this cone will be below the liquid state line of the M1 = 80 grams cone and parallel to it.

The plastic limit is given as

$$PL = LL - \frac{2\Delta w}{\log(M_2/M_1)}$$

Shrinkage Limit Test

The shrinkage limit is determined as follows. A mass of wet soil, m_1, is placed in a porcelain dish 44.5 mm in diameter and 12.5 mm high and then oven dried. With oven-dried soil still in the dish, the volume of shrinkage can be determined by filling the dish with mercury. The volume of mercury that fills the dish is equal to the shrinkage volume. The shrinkage limit is calculated from

$$SL = \frac{m_1 - m_2}{m_2} - \frac{(V_1 - V_2)\rho_w}{m_2}$$

where m_1 = mass of wet soil, m_2 = mass of oven-dried soil, V_1 = volume of wet soil, V_2 = volume of oven-dried soil, and ρ_w = density of water.

Direct shear test

The test is carried out on a soil sample confined in a metal box of square cross-section which is split horizontally at mid-height. A small clearance is maintained between the two halves of the box. The soil is sheared along a predetermined plane by moving the top half of the box relative to the bottom half. The box is usually square in plan of size 60 mm x 60 mm. A typical shear box is shown.

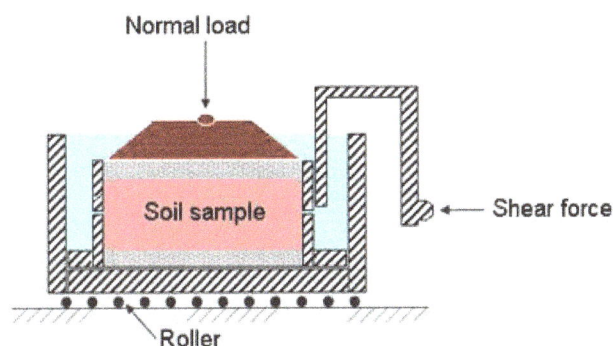

If the soil sample is fully or partially saturated, perforated metal plates and porous stones are placed below and above the sample to allow free drainage. If the sample is dry, solid metal plates are used. A load normal to the plane of shearing can be applied to the soil sample through the lid of the box.

Tests on sands and gravels can be performed quickly, and are usually performed dry as it is found that water does not significantly affect the drained strength. For clays, the rate of shearing must be chosen to prevent excess pore pressures building up.

As a vertical normal load is applied to the sample, shear stress is gradually applied horizontally, by causing the two halves of the box to move relative to each other. The shear load is measured together with the corresponding shear displacement. The change of thickness of the sample is also measured.

A number of samples of the soil are tested each under different vertical loads and the value of shear stress at failure is plotted against the normal stress for each test. Provided there is no excess pore water pressure in the soil, the total and effective stresses will be identical. From the stresses at failure, the failure envelope can be obtained.

The Test has Several Advantages

It is easy to test sands and gravels.

Large samples can be tested in large shear boxes, as small samples can give misleading results due to imperfections such as fractures and fissures, or may not be truly representative.

Samples can be sheared along predetermined planes, when the shear strength along fissures or other selected planes are needed.

The Disadvantages of the Test Include

- The failure plane is always horizontal in the test, and this may not be the weakest plane in the sample. Failure of the soil occurs progressively from the edges towards the center of the sample.

- There is no provision for measuring pore water pressure in the shear box and so it is not possible to determine effective stresses from undrained tests.

- The shear box apparatus cannot give reliable undrained strengths because it is impossible to prevent localized drainage away from the shear plane.

Triaxial Shear Test

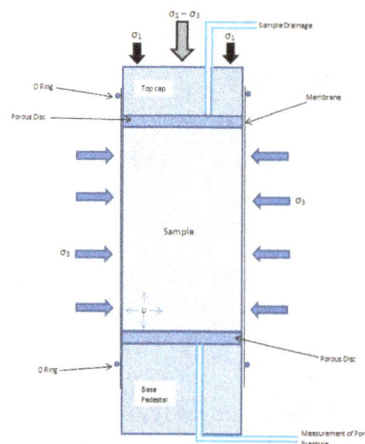

Figure: Stress Conditions in a Typical Triaxial Test

A typical triaxial test involves confining a cylindrical soil or rock specimen in a pressurized cell to simulate a stress condition and then shearing to failure, in order to determine the shear strength properties of the sample. Most triaxial tests are performed on high quality undisturbed specimens. The samples normally range from 38 mm to 100 mm samples, although samples considerably larger can be tested with the correct equipment. The test specimen most commonly has a height to diameter ratio of 2:1.

The sample will usually be saturated, then consolidated and finally sheared, most commonly only in compression - but extension tests may be undertaken with the correct equipment.

During the test the sample is subjected to stress conditions that attempt to simulate the in-situ stresses. Figure shows the stress conditions applied to a sample during a typical test; Figure shows the basic principles on effective stresses.

To summarize for a triaxial compression test:

σ_1 - Vertical (axial) Stress (think of this as the vertical load applied to the sample)

This also known as the Major Principle Stress

Can also be call σ_v

σ_3 - Confining Pressure (think of this as cell pressure)

This is also known as the Minor Principle Stress

Can also be called σ_h

U - Pore Pressure

Also known as U_w (Pore Water Pressure (P.W.P))

$\sigma_1 - \sigma_3$ – Deviator Stress (the stress due to the axial load applied to the specimen in excess of the confining pressure).

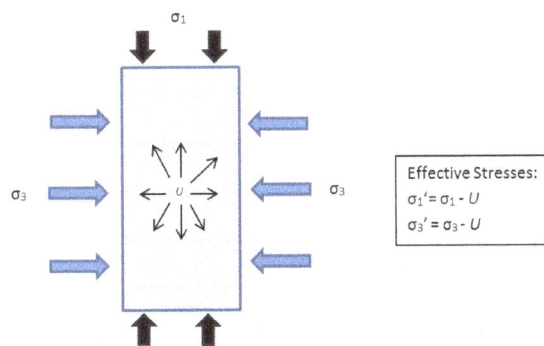

Effective Stresses:

$\sigma_1' = \sigma_1 - U$

$\sigma_3' = \sigma_3 - U$

Figure. Effective Stresses

σ_1' – Effective Vertical (axial) Stress

σ_3' – Effective Confining Pressure

U - Pore Pressure

Triaxial tests are one of the most widely performed tests in a geotechnical laboratory. The advantages of the test over other test methods used in the geotechnical laboratory used to determine shear strength (such as direct shear) is that specimen drainage can be controlled and pore pressure can be measured. The triaxial test enables parameters such as cohesion (c'), internal angle of friction (φ') and shear strength to be determined.

The triaxial test can also be used to determine other variables such as stiffness and permeability with the correct equipment.

A list of some of the common engineering issues that triaxial tests can be used for can be seen in table.

Table: Triaxial Tests used in common Engineering Problems

Field Problem	Type of Analysis	Type of Test
First Time Slope Failure	Effective Stress	CU or CD Triaxial
Cut Slope Failure	Effective Stress	CU Triaxial
Earth Dams	Total Stress	UU triaxial
	Effective Stress	CU Triaxial
		Triaxial permeability
Tunnel Linings	Total Stress	UU triaxial
	Effective Stress	CU Triaxial

It is common for UU triaxial tests to be used for short term engineering problems and CU and CD triaxials to look at long term engineering problems.

There are 3 Common Types of Triaxial Test

- UU Triaxial - Unconsolidated Undrained (this can also be termed QU for Quick Undrained)

- CU Triaxial - Consolidated Undrained triaxial

- CD Triaxial - Consolidated Drained triaxial

UU triaxial tests are termed total stress tests and CU/CD triaxial tests are termed effective stress tests. Figure shows the stresses involved with a total stress test and other figure summaries the stresses in an effective stress test.

CU and CD Triaxial tests normally start with a saturation stage, which then leads onto a consolidation stage. The consolidated stage can be either;

a) isotropic; the stress surrounding the sample is equal in all directions or

b) anisotropic; the stress surrounding the sample is not equal in all directions.

The tests are commonly abbreviated to CIU (Consolidated Isotropic Undrained) or CAU (Consolidated Anisotropic Undrained). In the last stage the sample is sheared to failure.

UU triaxial tests commonly do not have a saturation or consolidation stage performed; the test normally only consists of a shear stage. UU triaxial tests are quick tests, taking under 30 minutes to perform. The CU and CD triaxial tests take significantly longer to perform and can take weeks or even months to complete. The CD Triaxial commonly has the longest test duration, as during the shear stage of the test, pore pressure is not allowed to build up, resulting in very low strain rates. CU triaixals are quicker to perform as PWP is allowed to increase and the excess PWP is measured throughout the shear stage.

An additional test called unconfined compressive strength (UCS) can also be undertaken with a triaxial system. This shears a specimen under axial load with no confining pressure.

Typical Triaxial System – Typical Triaxial Cell

A typical triaxial cell is shown in Figure of this document showing a sample setup for testing. The setup shown is used for CU and CD triaxial tests. UU triaxial test can also be undertaken with this system or a simpler cheaper system can be purchased if only UU triaxial tests are required.

The sample is sealed inside a rubber membrane and then surround with water (cell pressure). The cell pressure is then used to apply a stress to the sample (σ_3) Back pressure can be added to saturate the sample if required. The sample is axially loaded during shearing with the force measured by the load cell, the deformation measured by the displacement transducer and if required the pore pressure in the sample can be measured. Volume change of the sample is measure from the back pressure line using a volume change unit or an automatic pressure controller.

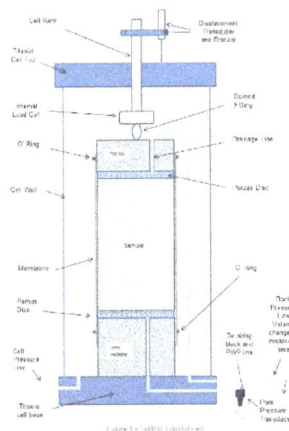

Figure Typical Triaxial cell

Typical Triaxial System – Compressed Air

A typical triaxial system for triaxials that uses compressed air can be seen in figure. This system consists of the following components that used for the following:

Pressure Panel – This enables water and compressed air to be correctly distributed to the right equipment to undertake a triaxial test. Cell Pressure (σ_3 - Confining Pressure) and Back pressure are controlled from here. A pressure gauge is included so that known pressures can be applied to the test specimen using the pressure regulators fitted to the panel.

Load Frame – The load frame is used to apply deformation to the triaxial specimen. The load frame can be controlled to a high level of accuracy (these tests normally require a slow speed). Load frames come in a range of capacities, VJ Tech can supply frames from 10kN to 250kN capacity. The one in Figure is a 50kN load frame and can apply loads of up to 50kN to a specimen.

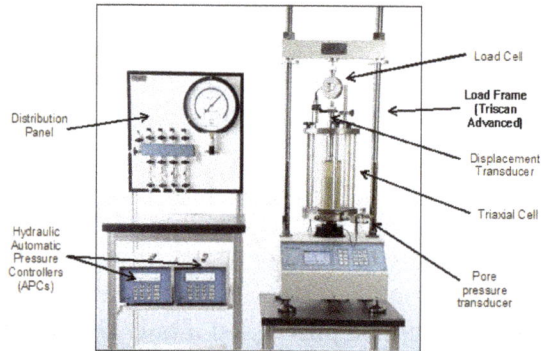

Figure Compressed Air triaxial system

Air/Water Cylinders – These provide an interface between compressed air and water in your triaxial system. They contain a rubber bladder which is filled with compressed air to pressurize the water inside the cylinder, this in turn provides the pressure used in the cell and back pressure for our triaxial test. The pressure the system can apply to samples is limited by the compressor.

Load Cell – The load cell or other load measuring device such as a load ring, provide the loads required to shear a triaxial specimen.

Displacement transducer – This could be a mechanical dial gauge, digital dial gauge or LSCT type transducer and is used to accurately measure the deformation applied to the triaxial specimen as it is sheared.

Triaxial Cell – The Triaxial cell comes in a range of sizes and pressure ratings. The cell contains the triaxial specimen and is pressurized throughout the test.

Pore Pressure Transducer – In CU and CD triaxial tests, the pressures inside the sample are measured (U - Pore Pressure). This is done using a pressure transducer attached to the base of the triaxial cell.

Automatic Volume Change unit – The automatic volume change unit is used in some triaxial tests to measure the volume of water going into the specimen and also the volume change of the specimen through the test.

With a system like this the pressures are set manually by the technician undertaking the test. It is common for the outputs from load cell, displacement transducers and pressure transducers to be data logged in some way and data analyzed using VJ Tech's Clisp Studio software.

Typical Triaxial System – Hydraulic Pressure Controller

A triaixal system for testing triaxials using hydraulic pressure can be seen in Figure below. The system consists of the following components:

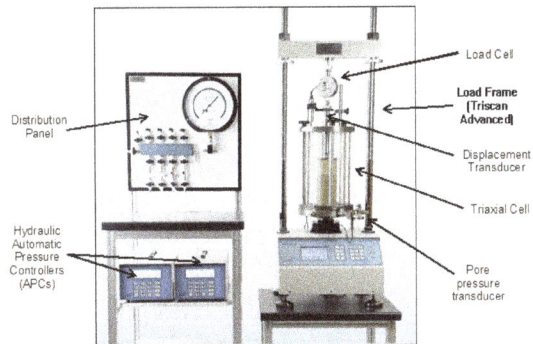

Figure: Hydraulic Pressure Controller Triaxial system

Distribution Panel – The distribution panel is used to connect the pressure controllers to the tri-axial cell. It also has the de-aired water system connected to it. It allows for the easy movement of water to the required location in the system without the need to disconnect lines. The pressure gauge on the panel is used for calibration, but is not used during testing.

Automatic Pressure/Volume Controllers – The automatic pressure and volume controllers (APCs) are used to generate pressure for the triaxial test. They use a stepper motor to pressurize a cylinder of water to generate cell pressure (σ_3 - Confining Pressure) and back pressure. No compressed air supply is required. The units are also able to measure volume change during the test. They are available in a range of capacities from 1mPa to 70mPa and a range of volumes from 200cc to 1000cc.

Load Frame – The load frame is used to apply deformation to the triaxial specimen. The load frame can be controlled to a high level of accuracy. Load frames come in a range of capacities, VJ Tech can supply frames from 10kN to 250kN capacity. The one in figure is a 50kN load frame and can apply loads of up to 50kN to a specimen. The load frame will commonly have a built-in data logger to log transducer data during the test; a separate data logger can be incorporated into the system if required.

Load Cell – The load cell or other load measuring device such as a load ring, provide the loads required to shear a triaxial specimen.

Displacement transducer – This could be a mechanical dial gauge, digital dial gauge or LSCT type transducer and is used to accurately measure the deformation applied to the triaxial specimen as it is sheared.

Triaxial Cell – The Triaxial cell comes in a range of sizes and pressure rating. The cell contains the triaxial specimen and is pressurized throughout the test.

Pore Pressure Transducer – In CU and CD triaxial tests the pressures inside the sample are measured (U - Pore Pressure), this is done using a pressure transducer attached to the base of the triaxial cell.

Additional Equipment Needed

For both CU/CD triaxial systems a de-aired water supply is also required for testing. The UU system can use a mains water supply if required.

Automatic systems require a computer and software package such as VJ Tech's Clisp Studio, to control the test and record transducer outputs.

A data logger (either inbuilt or external) will be used to store transducer readings taken during the test.

Sample preparation equipment will also be required, such as: sample extruder, sample moulds, membranes, 'o' rings, porous discs, ruler, balance and filter papers.

Test Procedure – Standards for Triaxial Tests

Triaxial tests are document in a range of international standards. These include:

British Standard 1377 part 7 and 8

ASTM D2850, D4767 and D2166

Australian standard AS1289 6.4.1 and 6.4.2

Hong Kong Geospec 3

German DIN 18137

Russian GOST 12248-96

Additional tests can also be performed with triaxial systems such as Stress Path analysis and anisotropic consolidation.

With additional equipment, a standard triaxial system can be upgraded to undertake the following tests:

Bender Element analysis (to determine the G_{max} of a sample).

Cyclic/Dynamic triaxial testing (to determine the cyclic shear strength of a sample and also used for modulus and damping properties).

Small Strain analysis (used for stiffness analysis) using on-sample transducers measuring axial and radial deformation can also be performed.

Test Procedure – System Preparation

Before any triaxial test is undertaken it is very important that checks are undertaken to verify the performance of the system. Failure to do this may mean either an extended test period or a sample being destroyed and no results obtained.

The following list is a suggested check list of the system before any test takes place. It is assumed the system has already been installed. Additional checks may be required depending on your local testing standard.

- Create enough de-aired water for the preparation and test (this can take a few hours to do).

- Clean the triaxial cell, paying special attention to the base and groove holding the 'o' ring.

- Clean 'o' ring in cell base and ensure it is free from defects.

- Flush the de-airing block that the pore pressure transducer is connected to. Pressurize and check the block for leaks.

- Check the pore pressure transducer calibration and recalibrate if necessary

- Flush the back pressure line.

- Check the back pressure line for leaks (following the procedure in most standards will take at least 24 hours).

- Check load cell and displacement transducers are accurate.

Test Procedure – Sample Setup

Most triaxial tests are performed on high quality undisturbed samples; local standards such as Euro code 7 give details on what is considered to be a high quality undisturbed sample; it is important to realize that sample disturbance (from collection, transport to lab and storage) will affect the results obtained from a triaxial test. Remoulded specimens can be tested and preparation procedures can be found in relevant standards and texts on this subject.

Samples are normally prepared to give a sample height to diameter ratio of 2:1. The sample ends are trimmed to ensure they are level and flat. This is commonly and most easily done using a sample mould or former for the diameter of sample that is being tested and a straight edge.

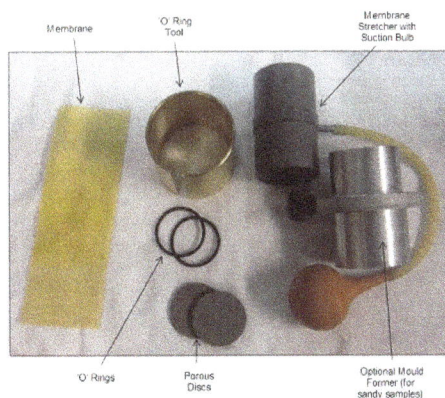

Figure Sample preparation equipment

The sample is then weighed so the bulk density can be determined and measured (both length and diameter) so the volume and area can be calculated. It is critical that the sample dimensions are accurate so that stress and strain being applied to the specimen during testing can be calculated accurately.

A number of critical checks should be made during this process:

1. Use a new membrane and make sure there are no holes in the membrane.

2. Ensure the porous discs are clean and not clogged with soil particles.

3. Place the 'o' rings using an 'o' ring tool, this will minimize sample disturbance.

4. Clean any loose material from the sample as you go. This loose material can cause leaks.

5. Take care not to hit the sample when placing the top of the cell over the sample.

During this preparation process care must be taken to limit sample disturbance. Sample disturbance can significantly affect the results obtained from the test.

Test Procedure – Saturation

Effective stress triaxials (CU and CD tests) require that the sample is saturated for testing. The reason for this is so that reliable measurements of pore pressure can be made. This is made possible by removing the air from the voids inside the sample.

The saturation process can be performed in a number of ways details of which can be found in the available Standards and text books. The most common though is incremental increase of cell and back pressure into the sample.

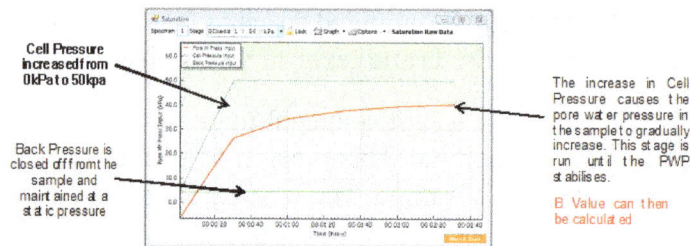

Figure: B Check Example

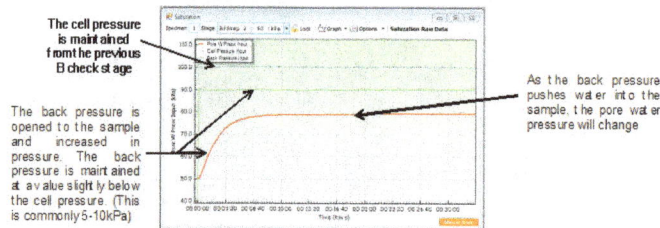

Figure: Back Pressure Step Example

This process gradually increases both the cell and back pressure so to dissolve any air that is in the voids of the soil sample. During this process the pore pressure coefficient B is calculated to determine the degree of saturation.

Figure shows the process that happens during a B check to determine the level of saturation. Most standards state that a B value greater than 0.95 indicates that a sample is sufficiently saturated.

The B value is calculated using the equation in figure.

$$B = \frac{\Delta U}{\Delta \sigma_3}$$

Figure: B Value Equation

ΔU = *Change in pore pressure*

$\Delta \sigma_3$ = Change in Confining Pressure

Figure shows the process that happens during a back pressure step. This step pushes water into the sample to try and saturate it if the B check shows the sample isn't saturated. These steps are repeated until the sample is saturated.

Test Procedure – Consolidation

After saturation, the specimen is normally consolidated to a stress condition representative of its in-situ condition. The consolidation is normally isotropic (the stresses applied vertically and horizontally to the sample are the same).

In effective stress tests the sample will be consolidated to an effective pressure. This is the difference between the confining (cell) pressure and the back pressure. If this is compared to the equation in Figure defining effective stress you will see this is incorrect.

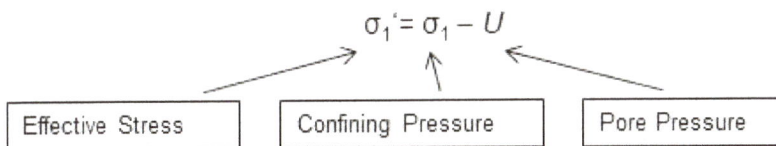

$$\sigma_1' = \sigma_1 - U$$

| Effective Stress | Confining Pressure | Pore Pressure |

At this point in the test pore pressure is replaced by back pressure to define our effective stress, to help simplify setup of this phase of the test. When the confining pressure and back pressure have been set to the correct values, there will be an excess of pore pressure in the sample (the pore pressure will be higher than the back pressure).

Once the consolidation is started the excess pore pressure in the sample will start to dissipate as the consolidation process drains of water from the sample decreasing its volume. Once this process is complete (the sample has stopped changing in volume) the pore pressure will be at or very near the back pressure level and pore pressure will be used to calculate the effective stress conditions the sample is under.

100% dissipation of excess pore pressure is not always possible so most standards suggest a minimum dissipation of 95% is achieved before the consolidation process is stopped.

Note during consolidation σ_3 is the same as σ_1.

Test Procedure – Shear

In the shear stage the axial force (σ_1) is gradually increased whilst the confining pressure (σ_3) is maintained until failure happens. This is the maximum shear force the soil can take.

This stage of the test is also commonly referred to as load stage, compression stage or shearing to failure stage.

During the shear stage the drainage conditions that are applied to the sample will determine if a drained or undrained test is performed. If the sample is not allowed to drain this is an undrained test, the sample will not change in volume during this test but will change in shape. If drainage is allowed, water will drain from the sample during the test, allowing the volume and the shape of the sample to change. In a CU (undrained) test, pore pressure changes (U) are allowed to take place and are measured. In a CD (Drained) test drainage is allowed to prevent pore pressure changes happening. In this type of test the volume change of the sample is measured during the shear stage.

From this stage of the test, as well as the shear strength being determined, the values of cohesion (c') and phi (φ') can be determined using Mohr circle and stress path plots.

Proctor Compaction Test

In the early 1930s, University of California, Berkeley student Ralph R. Proctor developed a method for determining the maximum density of soils. He established a laboratory procedure to define the moisture-density relationship of compacted, cohesive soils. Values from the Standard Proctor Test could be compared to unit weights and moistures of the same soils compacted in the field as structural earth fills to determine their degree of density and predict future performance. In 1958, the Modified Proctor Test was developed as an ASTM standard and it's still used today concurrently with the standard test.

Proctor Compaction Test Importance

Compaction is a method of mechanically increasing the density of soil, and it's especially valuable in construction applications. If this process is not performed properly, soil settlement can occur, resulting in unnecessary maintenance costs or failure of the pavement or structure.

Compacting soil

- Increases load capacity and stability

- Decreases permeability

- Prevents settlement of the soils or damage from frost

- Reduces water seepage, expansion and heaving

Proctor Tests

The Proctor Compaction Test and its variants are used to determine optimal moisture content for soils. This test is especially useful when determining the relationship between water content and the dry unit weight of soils to establish the maximum density of a soil needed for a fill area. The laboratory test serves a two-fold purpose by first determining the maximum density achievable for the materials in the field, as a reference. Secondly, it measures the effect moisture has on soil density. These values are often determined before earthwork begins to provide reference values for field testing.

Sample Preparation

A representative bulk sample is obtained for each type of material proposed for use in the earthwork operation. Back in the lab, processing of the samples begins with gradual air-drying to the desired moisture, usually around 10% or more below the anticipated optimum moisture. For cohesive soils, this can be expedited by breaking down clumps and spreading the sample out on open trays.

Once the soil is friable enough, the breakdown can continue more thoroughly. It is important to read and understand your particular test method carefully as there are a number of variables that can affect this stage of sample preparation. For most standard and modified Proctor variations, this means reducing the finer materials to pass through either a 4.75mm (#4) or 9.5mm (3/8in) sieve. Coarser materials are set aside for particle size determinations and in some cases for adding proportionally back into the final test specimens. At this stage, sample breakdown and coarse particle sizing are often performed concurrently.

Four or five specimens are prepared for the compaction points with increasing moisture contents and bracketing the estimated optimum water content. This requires some guesswork and experience is always helpful. Approximate optimum moisture of most cohesive soils can be estimated by manually squeezing a portion into a lump that will stick together, yet break cleanly into two sections when "bent". Weight of the specimens should be about 5lb (2.3kg) each for 4in (102mm) molds and 13lb (5.9kg) for 6in (152mm) molds to ensure enough compacted volume to properly fill the molds. Water is added incrementally to increase the individual moisture contents by about 2% for each specimen and mixed thoroughly. The prepared specimens are set aside in closed containers for a prescribed amount of standing time ranging up to 16 hours for proper moisture conditioning. The containers can be sealed metal cans, but heavy-duty zip-closure bags work well for this step.

Sample Compaction

For each Proctor point, the operator compacts the specimen into the pre-weighed empty Proctor Mold in three to five layers (lifts) according to the method required. Special Manual Rammers of 5.5lb (2.5kg) dropped from a 12in (305mm) height, or 10lb Rammers with 18in (457mm) drop height are used for compaction. Automatic Compactors are also available to make this process easier. The collar is removed and any excess is carefully trimmed with a straightedge tool so the compacted soil is flush with the top of the mold. Small voids can be manually filled with excess sample. The mold with sample is then weighed and recorded and the soil is extruded from the mold. A sample of the specimen is obtained to determine exact moisture content by oven drying, and the process is repeated for subsequent samples.

For each of the initial points, the mass and the unit weight of the soil will increase as the increased moisture lubricates particles and allows them to be consolidated into a denser state from the same compactive effort. By about the fourth point (if you're lucky and your estimates were correct) the mass of the sample will decrease as the volume of water reaches a point where it displaces soil particles in a given volume. This indicates that optimum moisture has been exceeded and having another point beyond that will make it a bit easier when constructing the final compaction curve.

Calculation and Plotting

The weight of each specimen is used to calculate wet unit weights and the oven-dried moistures are used to determine a dry unit weight for each point. The results are plotted on a graph as dry unit weight vs. moisture content and will show the curvilinear relationship that allows the maximum dry weight and optimum moisture for each type of soil to be established. These results are applied directly during field compaction tests to express the percent compaction of each test and determine if project design requirements are being met.

Equipment

- Soil Molds with either 4in or 6in diameters to hold compacted samples.

- Manual Soil Compactors enables compaction of an individual 4in Marshall asphalt sample into a stationary mold using a manually operated drop hammer.

- Mechanical Soil Compactors automatically count the number of hammer blows and shut off when a preset number is reached for improved accuracy, operation and reliability.

- Balance compliant with D4753 and 1-g readability to weigh the dry unit sample after compaction.

- Drying Oven that maintains uniformity to 230 ± 9°F (110 ± 5°C), to dry sample prior to testing.

- Straightedge to level and trim specimens held in the molds.

- E11 Sieves for particle size determinations.

- Pans for air drying, processing and mixing.

Tips

- Heavy duty plastic freezer bags are convenient for storage and moisture conditioning of individual compaction specimens.

- When preparing individual compaction specimens, it takes little time to prepare a couple of extra, in case your estimates are a bit off. Set up one on the drier side and one on the wet side to cover the bases.

- Some methods and materials require adding coarse material back into the final specimen. Set aside your plus size material as you break down the bulk sample. It's also convenient to sieve the coarse material now to reduce handling.

- Using a Sample Ejector expedites removal of compacted soil from the mold, increasing speed and efficiency and making it easier to obtain a representative moisture sample.

- Running five or six points of a proctor test can be physically demanding, especially with a Modified Proctor using a 10lb hammer to compact five layers.

Soil Respiration

Soil respiration is a measure of carbon dioxide (CO_2) released from the soil from decomposition of soil organic matter (SOM) by soil microbes and respiration from plant roots and soil fauna. It is an important indicator of soil health because it indicates the level of microbial activity, SOM content and its decomposition. In the short term high soil respiration rates are not always better; it may indicate an unstable system and loss of soil organic matter (SOM) because of excessive tillage, or other factors degrading soil health. It can be measured by simple methods or more sophisticated laboratory methods. The amount of soil respiration is an indicator of nutrients contained in organic matter being converted to forms available to crops (e.g., phosphate as PO_4, nitrate-nitrogen as NO_3, and sulfate as SO_4).

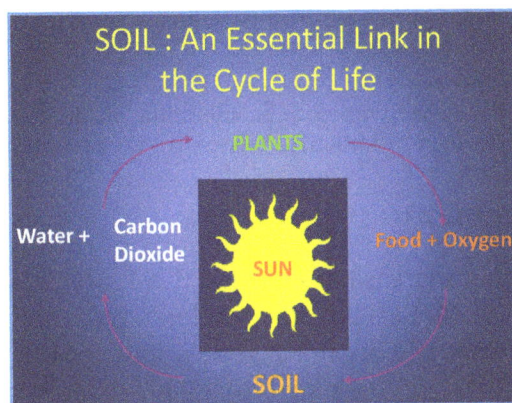

Figure: Soil microbial activities and respiration and mineralization of organic matter in soil completes the cycle of life on earth releasing water, oxidized minerals (NO_3, PO_4, etc.) and CO_2 needed by photosynthetic green plants to use the sun's energy to produce food (carbohydrates, etc.) and oxygen required for respiration to complete the cycle.

Inherent Factors Affecting Soil Respiration

Inherent factors that impact soil respiration, such as climate, cannot be changed. Inherent soil respiration rates depend on amount and quality of SOM, temperature, moisture, salinity, pH, and aeration. Biological activity of soil organisms varies seasonally, as well as daily. Microbial respiration more than doubles for every 10°C (18°F) soil temperatures rise up to a maximum of 35 to 40°C (95 to 104°F), beyond which soil temperature is too high, limiting plant growth, microbial activity and soil respiration.

Soil respiration increases with soil moisture up to the level where pores are filled with too much water limiting oxygen availability which interferes with soil organism's ability to respire (Figure). Ideal soil moisture is near field capacity, or when approximately 60 percent of pore space is filled with water. Respiration declines in dry soils due to the lack of moisture for microbes and other biological activity.

As soil water-filled pore space exceeds 80 percent, soil respiration declines to a minimum level and most aerobic microorganisms "switch tracks" and use nitrate (NO_3), instead of oxygen, resulting in loss of nitrogen, as nitrogen gases (N_2 and nitrogen oxides), emission of potent greenhouse gases, yield reduction, and increased N fertilizer expense.

Medium textured soils (silt and loam soils) are often favorable to soil respiration because of their good aeration, and high available water capacity. In clay soils, a sizeable amount of SOM is protected from decomposition by clay particles and other aggregates limiting soil respiration and associated mineralization (ammonification) of organic N. Sandy soils are typically low in SOM and have low available water capacity limiting soil respiration and N mineralization.

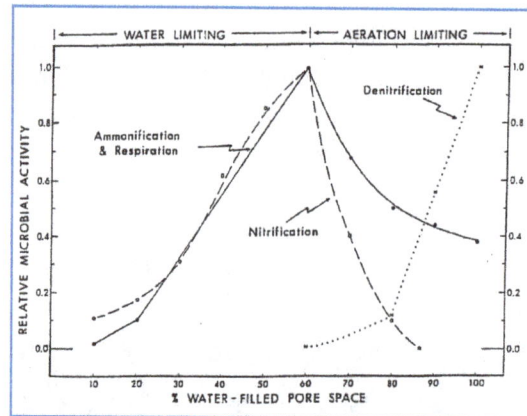

Figure: Relative aerobic (respiration, ammonification, nitrification) and anaerobic microbial activity (denitrification) as related to soil water-filled pore space.

Soil Respiration Management

Management practices can either increase or decrease SOM. Soil health and long term soil respiration improve with increased SOM. Leaving crop residues on the soil surface, use of no-till, use of cover crops, or other practices that add organic matter will increase soil respiration. Crop residues with a low carbon to nitrogen (C:N) ratio (e.g. soybean residue) decompose faster than residues with a high C:N ratio (e.g., wheat straw). High residue producing crops coupled with added N (from any source) increase decomposition and accrual of SOM. Conversely, tillage methods that remove, bury, or burn crop residues diminish SOM content will reduce soil respiration over the long term.

Irrigation in dry conditions and drainage of wet soils can significantly boost soil respiration. Soil respiration tends to be higher in crop rows than in between rows due to added contributions from plant roots. Compacted areas such as wheel tracks tend to have lower respiration than non-compacted areas because there is less aeration, less drainage, and higher water content.

Managing soil pH and salt content (salinity) is important because they regulate crop growth and nutrient availability and distribution which impact soil organisms responsible for SOM decomposition, and other processes contributing to soil respiration. Fertilizers may stimulate root growth and nourish microbes; however, at high concentrations, some fertilizers can become harmful to microbes responsible for soil respiration because of pH or salinity increases. Similarly, sludge or other organic materials with high concentrations of heavy metals, certain pesticides or fungicides, and salts may be toxic to microbial populations decreasing respiration.

Measures to improve SOM and/or soil porosity:

- Minimize soil disturbance and farm equipment activities when soils are wet,

- Use designated field roads or rows for equipment traffic,

- Reduce the number of trips across field,

- Subsoil to disrupt existing compacted layers,

- Cropping systems that include combinations of continuous no-till, cover crops, solid manure or compost application, diverse rotations with high residue crops and perennial legumes or grass used in rotation,

- Leave undisturbed residue on the soil surface rather than incorporating, burning, or removing.

Soil respiration rates respond to management measures such as plant residue or manure addition, tillage, and nitrogen applications as shown in Table. Temporary increases in soil respiration induced by certain management practices and have negative impact on SOM and long term soil respiration.

Table. Interpreting management impacts on soil respiration and soil organic matter (SOM).

Management Practice	Application	Short term Impacts	Long Term Impacts
Solid manure or organic material application	Provide additional carbon and N source for microbes to breakdown and increase biomass production.	Increased respiration when manure begins to breakdown and increased biomass production.	Positive impact on soil structure, fertility and SOM content.
High residue crops or cover crops used in rotation with high C:N ratio	High C:N ratio crops coupled with added N (from any source) increase decomposition and accrual of SOM.	High C:N ratio crop residue tie up nitrogen temporarily in order to break down residue, increased soil moisture, decreased erosion.	Positive impact on long term soil quality, fertility and SOM content.
Tillage such as annual disking, plowing etc.	Stirs the soil providing a temporary increase in oxygen for microbes to break down carbon sources	Provides a flush of nitrogen, other nutrients and CO_2 release immediately after tillage. Increases erosion rates, decomposition rate of residue, and other carbon sources.	Declines in SOM, soil quality, soil fertility.
Crop residue management	Leave residue on the surface increasing ground cover to protect the soil.	Increased crop residue cover can tie up nitrogen temporarily in order to break down residue, increased soil moisture, decreased erosion and cooler soil temperatures.	Positive impact on long term soil quality, fertility and SOM content.
Nitrogen fertilizer or manure application	Provides nitrogen (energy) source for microbes to break down high C:N ratio residue (e.g. corn stalks, wheat straw) quicker.	Temporary increase in respiration due to increased rate of breakdown of organic materials.	When managed correctly has an overall positive impact on SOM and soil quality by increasing production levels, and residue amounts.
Vehicle or farm equipment traffic	Compacts soil decreasing pore space, water movement, oxygen for microbes, and N loss from denitrification.	Decreases respiration, yields, water infiltration and increases runoff.	Production declines, increased soil erosion and runoff, decreased soil quality, compacted soils and reduced microbial activity.

Problems Related to Soil Respiration and Relationship to Soil Function

Soil respiration reflects the capacity of soil to sustain plant growth, soil fauna, and microorganisms. It indicates the level of microbial activity and SOM content and its decomposition. Soil respiration can be used to estimate nutrient cycling in the soil and the soil's ability to sustain plant growth.

Excessive respiration and SOM decomposition usually occurs after tillage due to destruction of soil aggregates and increased soil aeration. This depletes SOM, limits nutrient availability, and reduces yields.

Low soil respiration rates indicate that there is little or no SOM, or soil microbial activity. It may also signify that soil conditions (soil temperature, moisture, aeration, available N) are limiting biological activity and SOM decomposition. With low soil respiration, nutrients are not released from SOM to feed plants and soil organisms. Reduced soil respiration occurs when soils are flooded or saturated, and nitrogen is lost through denitrification and sulfur lost through volatilization.

Measuring Soil Respiration

Step by Step Procedure

1. Soil Sampling (Field): Soil respiration is variable both spatially and seasonally, and is strongly affected by organic matter, manure applications, oxygen levels, soil moisture, salinity (EC) and soil temperature. Soil is sampled in a fresh condition just before the test is performed. At least 10 small samples should be gathered randomly from an area that represents soil type and management history with a probe from the surface 0-6 inch depth and placed in the small plastic bucket. Repeat this for each sampling area.

2. Mixing (Field): Mix soil in the plastic bucket just well enough to be homogeneous and remove roots, residue, large stones and residues from sample and place in a labeled plastic zip bag.

3. Add Water If Needed (Field or Classroom): The sample should have ideal moisture (near field capacity) for growing conditions. If field conditions are dry it is best to add water 24 prior to sampling.

Bulk Density

Bulk density is an indicator of soil compaction. It is calculated as the dry weight of soil divided by its volume. This volume includes the volume of soil particles and the volume of pores among soil particles. Bulk density is typically expressed in g/cm^3.

Bulk density reflects the soil's ability to function for structural support, water and solute movement, and soil aeration. Bulk densities above thresholds indicate impaired function. Bulk density is also used to convert between weight and volume of soil. It is used to express soil physical, chemical and biological measurements on a volumetric basis for soil quality assessment and comparisons between management systems. This increases the validity of comparisons by removing error associated with differences in soil density at time of sampling.

Table: General relationship of soil bulk density to root growth based on soil texture.

Soil Texture	Ideal bulk densities for plant growth (g/cm3)	Bulk densities that restrict root growth (g/cm3)
Sandy	< 1.60	> 1.80
Silty	< 1.40	> 1.65
Clayey	< 1.10	> 1.47

Specific problems that might be caused by poor function: High bulk density is an indicator of low soil porosity and soil compaction. It may cause restrictions to root growth, and poor movement of air and water through the soil. Compaction can result in shallow plant rooting and poor plant growth, influencing crop yield and reducing vegetative cover available to protect soil from erosion. By reducing water infiltration into the soil, compaction can lead to increased runoff and erosion from sloping land or waterlogged soils in flatter areas. In general, some soil compaction to restrict water movement through the soil profile is beneficial under arid conditions, but under humid conditions compaction decreases yields.

The following practices can lead to poor bulk density:

- Consistently plowing or disking to the same depth,

- Allowing equipment traffic, especially on wet soil,

- Using a limited crop rotation without variability in root structure or rooting depth,

- Incorporating, burning, or removing crop residues,

- Overgrazing forage plants, and allowing development of livestock loafing areas and trails, and

- Using heavy equipment for building site preparation or land smoothing and leveling.

Any practice that improves soil structure decreases bulk density; however, in some cases these improvements may only be temporary. For example, tillage at the beginning of the growing season temporarily decreases bulk density and disturbs compacted soil layers, but subsequent trips across the field by farm equipment, rainfall events, animals, and other disturbance activities can recompact soil.

On cropland, long-term solutions to bulk density and soil compaction problems revolve around decreasing soil disturbance and increasing soil organic matter. A system that uses cover crops, crop residues, perennial sod, and/or reduced tillage results in increased soil organic matter, less disturbance and reduced bulk density. Additionally, the use of multi-crop systems involving plants with different rooting depths can help break up compacted soil layers.

To reduce the likelihood of high bulk density and compaction:

- Minimize soil disturbance and production activities when soils are wet,

- Use designated field roads or rows for equipment traffic,

- Reduce the number of trips across the area,

- Subsoil to disrupt existing compacted layers, and

- Use practices that maintain or increase soil organic matter.

Grazing systems that minimize livestock traffic and loafing, provide protected heavy use areas, and adhere to recommended minimum grazing heights reduce bulk density by preventing compaction and providing soil cover.

Conservation practices resulting in bulk density favorable to soil function include:

- Conservation Crop Rotation

- Cover Crop

- Deep Tillage

- Prescribed Grazing

- Residue and Tillage Management

References

- Triaxial-testing-an-introduction: vjtech.co.uk, Retrieved 12 July 2018

- Proctor-compaction-test-a-basic-guide: globalgilson.com, Retrieved 26 April 2018

- Bulk-density, indicators: soilquality.org, Retrieved 20 May 2018

- Consistency-soil-atterberg-limits, geotechnical-engineering: mathalino.com, Retrieved 14 June 2018

- Introduction-to-soils-soil-quality: extension.psu.edu, Retrieved 09 March 2018

Chapter 8

Soil Organic Matter

The organic matter content of the soil that consists of animal and plant residues at all their stages of decomposition, soil organisms along with their cells and tissues together is known as soil organic matter. The topics covered in this chapter such as, soil organic matter component and management of soil organic matter are crucial for a deeper understanding of soil.

Soil organic matter is the fraction of the soil that consists of plant or animal tissue in various stages of breakdown (decomposition). Most of our productive agricultural soils have between 3 and 6% organic matter.

Soil organic matter contributes to soil productivity in many different ways.

Organic matter is made up of different components that can be grouped into three major types:

1. Plant residues and living microbial biomass.

2. Active soil organic matter also referred to as detritus.

3. Stable soil organic matter, often referred to as humus.

The living microbial biomass includes the microorganisms responsible for decomposition (breakdown) of both plant residues and active soil organic matter or detritus. Humus is the stable fraction of the soil organic matter that is formed from decomposed plant and animal tissue. It is the final product of decomposition.

The first two types of organic matter contribute to soil fertility because the breakdown of these fractions results in the release of plant nutrients such as nitrogen, phosphorus, potassium, etc.

The humus fraction has less influence on soil fertility because it is the final product of decomposition (hence the term "stable organic matter"). However, it is still important for soil fertility management because it contributes to soil structure, soil tilth, and cation exchange capacity. This is also the fraction that darkens the soil's color.

Benefits of Stable Soil Organic Matter

There are numerous benefits to having a relatively high stable organic matter level in an agricultural soil. These benefits can be grouped into three categories:

Physical Benefits

- Enhances aggregate stability, improving water infiltration and soil aeration, reducing runoff.
- Improves water holding capacity.

- Reduces the stickiness of clay soils making them easier to till.

- Reduces surface crusting, facilitating seedbed preparation.

Chemical Benefits

- Increases the soil's CEC or its ability to hold onto and supply over time essential nutrients such as calcium, magnesium and potassium.

- Improves the ability of a soil to resist pH change; this is also known as buffering capacity.

- Accelerates decomposition of soil minerals over time, making the nutrients in the minerals available for plant uptake.

Biological Benefits

- Provides food for the living organisms in the soil.

- Enhances soil microbial biodiversity and activity which can help in the suppression of diseases and pests.

- Enhances pore space through the actions of soil microorganisms. This helps to increase infiltration and reduce runoff.

Organic Materials

Over time, the application and incorporation of organic materials can result in an increase in stable soil organic matter levels. Sources of organic materials include:

- Crop residues.

- Animal manure.

- Compost.

- Cover crops (green manure)

- Perennial grasses and legumes.

The quickest increases are obtained with sources that are high in carbon such as compost or semi-solid manure.

Figure: Compost application can increase soil organic matter levels over time.

Organic Matter Management

Farm practices that help to maintain or increase soil organic matter levels:

- Use of conservation tillage practices (for example zone tillage or no-till). Tillage exposes the organic matter to air and will result in the lowering of stable organic matter due to increased mineralization rates and erosion losses.

- Rotation of annual row crops with perennial grass or legume sods will reduce erosion and build up organic matter as a result of the decomposition of the root mass.

- Establishment of legume cover crops will enhance organic matter accumulation by providing the nitrogen (N) needed for decomposition of freshly added organic materials, especially those with a high C to N ratio (corn stover, cereal straw, heavily bedded manure, etc.).

- Avoiding soil compaction which increases waterlogging, and maintaining proper pH to enhance microbial activity and decomposition of freshly added materials.

Actual buildup of stable organic matter will, in addition to the amount and source of organic material added, and tillage and rotation practices, also depend on:

- Soil temperature.

- Precipitation and soil moisture holding capacity.

- Soil type and drainage class.

- Existing microbial community.

- Soil fertility status and soil pH.

Monitoring Soil Organic Matter

To get an idea of the effect of farm management practices on soil organic matter buildup or decrease, soil samples should be taken over time. Consistency in sampling time is important to build records for fields over time. Although other tests are available, most laboratories will do a loss-on-ignition (LOI) test to estimate the organic matter content of the soil. At Cornell University, soil is exposed to 1050 C (2210 F) for 1.5 hours to remove soil moisture and then to 5000 C (9320 F) for 2 hours to determine LOI. Not all laboratories us the same method so for accurate records over time, it is important to consistently use the same laboratory service.

With careful management the preservation and accumulation of soil organic matter can help to improve soil productivity resulting in greater farm profitability.

Soil Organic Matter Component

SOM consists of a heterogeneous mixture of organic matter (OM) originating from plant (the major source), microbial (a minor source), and animal (a minor source) residues, exhibiting different stages of biological oxidation. A substantial fraction of SOM may be associated with mineral surfaces and protected against microbial enzyme attack, which complicates the study of SOM composition and dynamics by hindering extractability and measurability. Therefore, SOM remains largely uncharacterized at the molecular-level. However, structural information can be obtained on various SOM components that are amenable to advanced analytical techniques. Many structurally unique biochemical ('biomarkers') carry the information of their origins and/or environmental settings, and hence can be used to analyze the source and/or degradation stage of SOM. Some common SOM biomarkers include plant wax lipids, cutin- or suberin-derived lipids, lignin-derived phenols, and microbial phospholipid fatty acids.

Soil Lipids

Soil lipids, operationally defined as a heterogeneous group of organic substances that are insoluble in water but extractable with non-polar solvents, are common constituents of SOM. They range from simple structures such as alkanes, alkanols, alkanoic acids, and steroids to complex unknown lipids. Soil lipids bear important information about SOM inputs and influence the surface properties of aggregates and contaminant sorption. Soil lipids are primarily plant-derived and include plant waxes and biopolymers (such as suberin in the peridermis of barks and roots and cutin in leaf cuticles). Microorganisms are minor contributers to soil lipids, such as branched short-chain ($<C_{20}$) alkanoic acids, hopanoids, and ergosterol. Based on their structure and transformation processes in the soil, soil lipids can be readily extracted by organic solvents (solvent-extractable lipids), be bound to SOM by ester-linkages (bound lipids), or be non-extractable even through mild chemolytic methods such as catalyzed hydrolysis.

Solvent-extractable (free) lipids usually comprise less than 10% of SOM and they contain characteristic biomarkers that can give information about the source and degradation stage of residues in SOM. For instance, even-numbered long-chain ($>C_{20}$) alkanoic acids are common constituents of plant wax lipids whereas the degradation stage of plant-derived steroids may be assessed by comparing the ratio of plant steroids (β-sitosterol, stigmasterol, and stigmastanol) to their degradation products (stimasta-3,5-dien-7-one and sitosterone) in the soil.

By comparison, ester-bound lipids are not extractable with organic solvents, but they can be cleaved from SOM using chemolytic methods such as alkaline hydrolysis. The predominant long-chain ω-hydroxyalkanoic and α,ω-alkanedioic acids are typical biomarkers for suberin, primarily indicating root or bark inputs into the soil, while C_{16} and C_{18} ω-hydroxyalkanoic acids with mid-chain hydroxy or epoxy groups are biomarkers for cutin or leaf inputs. Suberin and cutin biopolymers are the major sources of bound (hydrolysable) aliphatic lipids in SOM, but not much is known about the degradation of suberin and cutin in soils. Bound lipids are considered to be less prone to microbial attack through chemical bonding and thus more stable than solvent-extractable lipids. Recent studies have shown that recalcitrant aliphatic SOM components have the potential to promote carbon sequestration in soils. It is hence important to examine the transformation and preservation of bound lipids in the soil under a changing climate.

Non-hydrolysable soil lipids are the lipid fraction that is not extractable by mild chemolytic methods (solvent extraction and hydrolysis) and are the least chemically understood. Non-hydrolysable SOM is considered to be a major part of the stable soil carbon pool, and mineral protection or chemical transformation (such as cross-linking) may contribute to its recalcitrance. Alternatively, non-hydrolysable soil lipids may include inputs from cutan (the non-hydrolysable and non-extractable biopolymer in leaf cuticles) and suberan.

Lignin-Derived Phenols

Lignin is the second most abundant biopolymer (after cellulose and hemicellulose) in nature and a large contributor to SOM. So far, there is no method for analyzing or quantifying the lignin macromolecules directly in soil. However, lignin-derived phenols or monomers (vanillyls, syringyls, and cinnamlyls; Figure) can be released from the macromolecular matrix of the biopolymer or soils by chemolytic methods such as CuO oxidation. The composition of lignin monomers is characteristic for major plant groups (angiosperms, gymnosperms) and is commonly used to describe the major plant sources. Ratios of lignin-derived phenolic acids to their corresponding aldehydes (Ad/Al) are useful tools for determining the stage of lignin degradation in soils and sediments. Lignin in SOM is usually considered to be refractory due to its aromaticity and slow decomposition rates in litters. Only certain groups of fungi (white-rot and brown-rot fungi) are able to efficiently biodegrade lignin in terrestrial environments. It is therefore important to monitor lignin degradation with environmental changes and with potential shifts in the microbial community structure brought on by environmental changes.

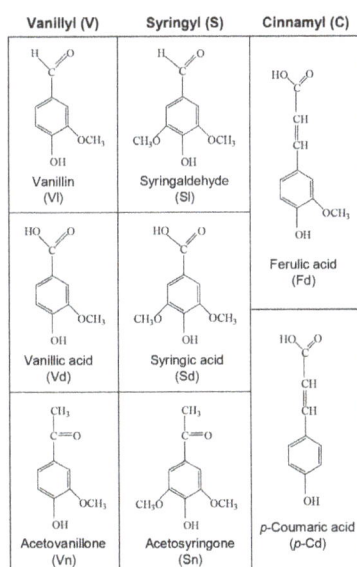

Figure: Structures of lignin-derived phenols (monomers) isolated by CuO oxidation.

Microbial Phospholipid Fatty Acids (PLFAs)

Compared with plant inputs, microbial biomass represents a minor (1~5%) yet arguably the most active fraction of SOM. Characteristic biomarkers of microbial biomass include ergosterol (from fungi), amino sugars (such as glucosamine and muramic acid), branched short-chain alkanoic acids, 3-hydroxyalkanoic acids, hopanoids, and PLFAs. Among them, PLFAs are only found in viable cells and hence are characteristic biomarkers for living microorganisms. Based on their chemical structures, such as branching within the molecule or the occurrence of double bonds, various PLFAs can be used to establish the notional proportions of fungi, Gram-positive bacteria (including actinomycetes) or Gram-negative bacteria, and hence to characterize microbial community structure in the soil.

Other SOM Components

It is worth noting that the SOM biomarkers mentioned previously are of significant importance in investigating SOM structural alterations under global changes such as global warming and elevated CO_2. Soil lipids and lignin-derived phenols provide insights into the inputs and degradation of plant-derived aliphatic molecules and lignin in the soil respectively, which may contribute to the sequestration of soil carbon in the long term, whereas PLFAs yield information on the living microbial biomass and community composition. By comparison, cellulose and hemicellulose, the most abundant biopolymers on earth, are efficiently degraded by fungi and bacteria in the aerobic litter layer and hence found in only low amounts in mineral soils. Similarly, non-cellulosic carbohydrates and proteins that occur in plants as well as microorganisms are either not source-specific or biochemically labile, such that they are not very useful indicators of SOM degradation or do not have large carbon sequestration potentials in the long term. Finally, the identified biomarkers only represent a small fraction of SOM. The majority of SOM remains uncharacterized at the molecular-level because they have undergone extensive transformations during humification processes or are closely associated with mineral matrix and/or present in macromolecules.

Management of Soil Organic Matter

There are four general strategies for organic matter management. First, use crop residues more effectively and find new sources of residues to add to soils. New residues can include those you grow on the farm, such as cover crops, or those available from various local sources. Second, try to use a number of different types of materials—crop residues, manures, composts, cover crops, leaves, etc. It is important to provide varied residue sources to help develop and maintain a diverse group of soil organisms. Third, although use of organic materials from off farm can be a good source for building soil organic matter and adding nutrients, some farmers overload their fields with excess nutrients by excess imports of organic materials. Crop residues (including cover crops) as well as on-farm-derived animal manures and composts help to supply organic materials and cycle nutrients without a buildup of excessive levels of nutrients. Fourth, implement practices that decrease the loss of organic matter from soils because of accelerated decomposition or erosion.

All practices that help to build organic matter levels either add more organic materials than in the past or decrease the rate of organic matter loss from soils. In addition, practices to build organic matter will usually enhance beneficial organisms and/or stress pests (table). Those practices that do both may be especially useful. Practices that reduce losses of organic matter either slow down the rate of decomposition or decrease the amount of erosion. Soil erosion must be controlled to keep organic matter–enriched topsoil in place. In addition, organic matter added to a soil must either match or exceed the rate of loss by decomposition. These additions can come from manures and composts brought from off the field, crop residues and mulches that remain following harvest, or cover crops. Reduced tillage lessens the rate of organic matter decomposition and also may result in less erosion. When reduced tillage increases crop growth and residues returned to soil, it is usually a result of better water infiltration and storage and less surface evaporation.

Using Organic Materials

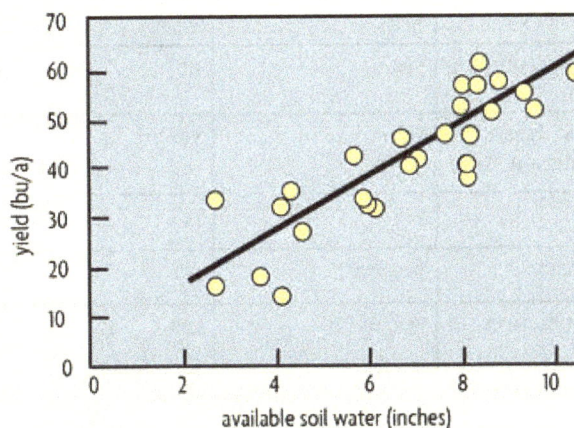

Figure Relationship between winter wheat grain yield and soil water
at wheat planting over six years. Modified from Nielsen et al.

Amounts of Crop Residues

Crop residues are usually the largest source of organic materials available to farmers. The amount of crop residue left after harvest varies depending on the crop. Soybeans, potatoes, lettuce, and corn silage leave little residue. Small grains, on the other hand, leave more residues, while sorghum and corn harvested for grain leave the most. A ton or more of crop residues per acre may sound like a lot of organic material being returned to the soil. However, keep in mind that after residues are decomposed by soil organisms, only about 10–20% of the original amount is converted into stable humus.

The amount of roots remaining after harvest also can range from very low to fairly high. In addition to the actual roots left at the end of the season, there are considerable amounts of sloughed-off root cells, as well as exudates from the roots during the season. This may actually increase the plant's belowground inputs of organic matter by another 50%. Probably the most effective way to increase soil organic matter is to grow crops with large root systems. Compared to aboveground residues, the organic material from roots decomposes more slowly, contributes more to stable soil organic matter, and, of course, does not have to be incorporated into the soil to achieve deep distribution. When no-till is used, root residues, along with root exudates given off when they were alive, tend to promote formation and stabilization of aggregates—more so than surface derived residue. One of the reasons that the many soils of the Midwest are so rich is that for thousands of years prairie plants with extensive and deep root systems grew there—annually contributing large quantities of organic matter deep into the soil.

TABLE: Effects of Different Management Practices on Gains and Losses of Organic Matter, Beneficial Organisms, and Pests

Management Practice	Gains increase	Losses decrease	Enhance beneficials (EB), Stress Pests (SP)
Add materials (manures, composts, other organic materials) from off the field	yes	no	EB, SP
Better utilize crop residue	yes	no	EB
Include high-residue-producing crops in rotation	yes	no	EB, SP
include sod crops (grass/legume forages) in rotation	yes	yes	EB, SP
Grow cover crops	yes	yes	EB, SP
Reduce tillage intensity	yes/no*	yes	EB
Use conservation practices to reduce erosion	yes/no*	yes	EB
*Practice may increase crop yields, resulting in more residue.			

Some farmers remove aboveground residues such as small grain straw from the field for use as animal bedding or to make compost. Later, these residues return to contribute to soil fertility as manures or composts. Sometimes residues are removed from fields to be used by other farmers or to make another product. There is increasing interest in using crop residues as a feedstock for the production of biofuels. This activity could cause considerable harm to soil health if sufficient residues are not allowed to return to soils.

Burning of wheat, rice, and other crop residues in the field still occurs, although it is becoming less common in the United States as well as in other countries. Residue is usually burned to help control insects or diseases or to make next year's fieldwork easier. Residue burning may be so widespread in a given area that it causes a local air pollution problem. Burning also diminishes the amount of organic matter returned to the soil and the amount of protection against raindrop impact.

Sometimes important needs for crop residues and manures may prevent their use in maintaining or building soil organic matter. For example, straw may be removed from a grain field to serve as mulch in a strawberry field. These trade-offs of organic materials can sometimes cause a severe soil-fertility problem if allowed to continue for a long time. This issue is of much more widespread importance in developing countries, where resources are scarce. In those countries, crop residues and manures frequently serve as fuel for cooking or heating when gas, coal, oil, and wood are not available. In addition, straw may be used in making bricks or used as thatch for housing or to make fences. Although it is completely understandable that people in resource-poor regions use residues for such purposes, the negative effects of these uses on soil productivity can be substantial. An important way to increase agricultural productivity in developing countries is to find alternate sources for fuel and building materials to replace the crop residues and manures traditionally used.

Table: Estimated Root Residue Produced by Crops

Crop	Estimated Root Residues (lbs/acre)
Native prairie	15,000-30,000
Italian ryegrass	2,600-4,500
Winter cereal	1,500-2,600
Red clover	2,200-2,600
Spring cereal	1,300-1,800
Corn	3,000-4,000
Soybeans	500-1,000
Cotton	500-900
Potatoes	300-600

Using Residues as Mulches

Crop residues or composts can be used as mulch on the soil surface. This occurs routinely in some re-duced-tillage systems when high-residue-yielding crops are grown or when killed cover crops remain on the surface. In some small-scale vegetable and berry farming, mulching is done by applying straw from off site. Strawberries grown in the colder, northern parts of the country are routinely mulched with straw for protection from winter heaving. The straw is blown on in late fall and is then moved into the interrows in the spring, providing a surface mulch during the growing season.

Mulching has Numerous Benefits, Including

- enhanced water availability to crops due to better infiltration into the soil and less evapora-tion from the soil (approximately 1/3 of water loss in dry land irrigated agriculture is from evaporation from the soil, which can be greatly reduced by using a surface mulch)

- weed control

- less extreme changes in soil temperature

- reduced splashing of soil onto leaves and fruits and vegetables (making them look better as well as reducing diseases)

- reduced infestations of certain pests (Colorado potato beetles on potatoes and tomatoes are less severe when these crops are grown in a mulch system)

On the other hand, residues mulches in cold climates can delay soil warming in the spring, reduce early-season growth, and increase problems with slugs during wet periods. When it is important to get a rotation crop in early, you might consider using a low-residue crop like soybeans the previous year. Of course, one of the reasons for the use of plastic mulches (clear and black) for crops like tomatoes and melons is to help warm the soil.

Residue Management in Arid and Semiarid Regions

In arid and semiarid regions water is usually the most common limitation to crop yields. For win-ter wheat in semiarid regions, for example, the available water at planting often foretells final yields (figure). Thus, in order to provide more available water for crops, we want to use practices that help store more water in soils and keep it from evaporating directly to the atmosphere. Standing residue allows more snow to be maintained in the field after being deposited, significantly increasing available soil water in spring— sunflower stalks used in this way can increase soil water by 4 to 5 inches. And a mulch during the growing season helps both to store water from irrigation or rainfall and to keep it from evaporating.

Effects of Residue Characteristics on Soil

Decomposition Rates and Effects on Aggregation

Residues of various crops and manures have different properties and, therefore, have different effects on soil organic matter. Materials with low amounts of harder-to-degrade hemicellulose, polyphenols, and lignin, such as cover crops (especially legumes) when still very green and soybean residue, decompose rapidly (figure) and have a shorter-term effect on soil organic matter levels than residues with high levels of these chemicals (for example, cornstalks and wheat straw). Manures, especially those that contain lots of bedding (high in hemicellulose, polyphenols, and lignin), decompose more slowly and tend to have more long-lasting effects on total soil organic matter than crop residues and manures without bedding. Also, cows—because they eat a diet containing lots of forages that are not completely decomposed during digestion—produce manure with longer-lasting effects on soils than non-ruminants, such as chickens and hogs, that are fed exclusively a high-grain and low-fiber diet. Composts contribute little active organic matter to soils but add a lot of well-decomposed materials (figure).

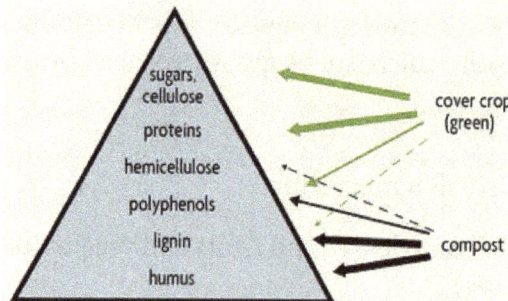

Figure Different types of residues have varying effects on soils (thicker lines indicate more material, dashed lines indicate very small percentages).

In general, residues containing a lot of cellulose and other easy-to-decompose materials will have a greater effect on soil aggregation than compost, which has already undergone decomposition. Because aggregates are formed from by-products of decomposition by soil organisms, organic additions like manures, cover crops, and straw will usually enhance aggregation more than compost. (However, adding compost does improve soils in many ways, including increasing the water-holding capacity).

Although it's important to have adequate amounts of organic matter in soil, that isn't enough. A variety of residues are needed to provide food to a diverse population of organisms, provide nutrients to plants, and furnish materials that promote aggregation. Residues low in hemicellulose and lignin usually have very high levels of plant nutrients. On the other hand, straw or sawdust (containing a lot of lignin) can be used to build up organic matter, but a severe nitrogen deficiency and an imbalance in soil microbial populations will occur unless a readily available source of nitrogen is added at the same time. In addition, when insufficient N is present, less of the organic material added to soils actually ends up as humus.

C:N Ratio of Organic Materials and Nitrogen Availability.

The ratio of the amount of a residue's carbon to the amount of its nitrogen influences nutrient availability and the rate of decomposition. The ratio, usually referred to as the C:N ratio, may vary from around 15:1 for young plants, to between 50:1 and 80:1 for the old straw of crop plants, to over 100:1 for sawdust. For comparison, the C:N ratio of soil organic matter is usually in the range of about 10:1 to 12:1, and the C:N of soil microorganisms is around 7:1. The C:N ratio of residues is really just another way of looking at the percentage of nitrogen (figure). A high C:N residue has a low percentage of nitrogen. Low C:N residues have relatively high percentages of nitrogen. Crop residues usually average 40% carbon, and this figure doesn't change much from plant to plant. On the other hand, nitrogen content varies greatly depending nitrogen release on the type of plant and its stage of growth.

If you want crops to grow immediately following the application of organic materials, care must be taken to make nitrogen available. Nitrogen availability from young, and very green plants, decompose rapidly in the soil and, in the process, may readily release plant nutrients. This could be compared to the effect of sugar eaten by humans, which results in a quick burst of energy. Some of the substances in older plants and in the woody portion of trees, such as lignin, decompose very slowly in soils. Materials such as sawdust and straw, mentioned above, contain little nitrogen. Well-composted organic residues also decompose slowly in the soil because they are fairly stable, having already undergone a significant amount of decomposition.

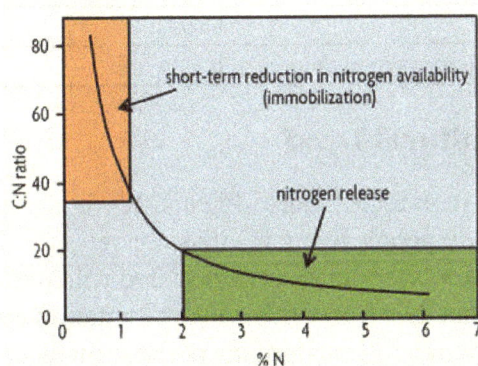

Figure Nitrogen release and immobilization with changing nitrogen content. Based on data Vigil and Kissel.

Mature plant stalks and sawdust that have C:N over 40:1 may cause temporary problems for plants. Microorganisms using materials that contain 1% nitrogen (or less) needs extra nitrogen for their growth and reproduction. They will take the needed nitrogen from the surrounding soil, diminishing the amount of nitrate and ammonium available for crop use. This reduction of soil nitrate and ammonium by microorganisms decomposing high C:N residues is called immobilization of nitrogen.

When microorganisms and plants compete for scarce nutrients, the microorganisms usually win, because they are so well distributed in the soil. Plant roots are in contact with only 1–2% of the soil volume, whereas microorganisms populate almost the entire soil. The length of time during which the nitrogen nutrition of plants is adversely affected by immobilization depends on the quantity of residues applied, their C:N ratio, and other factors influencing microorganisms, such as fertilization practices, soil temperature, and moisture conditions. If the C:N ratio of residues is in the teens or low 20s, corresponding to greater than 2% nitrogen, there is more nitrogen present than the microorganisms need for residue decomposition. When this happens, extra nitrogen becomes available to plants fairly quickly. Green manure crops and animal manures are in this group. Residues with C:N in the mid-20s to low 30s, corresponding to about 1–2% nitrogen, will not have much effect on short-term nitrogen immobilization or release.

Table: C:N Ratios of Selected Organic Materials

Material	C:N
Soil	10-12
Poultry manure	10
Clover and alfalfa (early)	13
Compost	15
Dairy manure (low bedding)	17
Alfalfa hay	20
Green rye	36
Corn stover	60
Wheat, oat, or rye straw	80
Oak leaves	90
Fresh sawdust	400
Newspaper	600
Note: Nitrogen is always 1 in the ratios.	

Sewage Sludge on Agricultural Land

In theory, using sewage sludge—commonly called bio solids—on agricultural land makes sense as a way to resolve problems related to people living in cities, far removed from the land that grows their food. However, there are some troublesome issues associated with agricultural use of sludges. By far, the most important problem is that they frequently contain contaminants from industry and from various products used around the home. Although many of these metal contaminants naturally occur at low levels in soils and plants, their high concentrations in some sludges create a potential hazard. The U.S. standards for toxic materials in sludges are much more lenient than those in some other industrialized countries and permit higher loading of potentially toxic metals. So, although you are allowed to use much sludge, you should carefully examine sludge's contents before applying it to your land.

Another issue is that sludges are produced by varied processes and, therefore, have different properties. Most sludges are around neutral pH, but, when added to soils; cause some degree of acidification, as do most nitrogen fertilizers. Because many of the problem metals are more soluble under acidic conditions, the pH of soils receiving these materials should be monitored and maintained at around 6.8 or above. On the other hand, lime (calcium hydroxide and ground limestone used together) is added to some sludges to raise the pH and kill disease bacteria. The resulting "lime-stabilized" sludge has extremely high levels of calcium, relative to potassium and magnesium. This type of sludge should be used primarily as a liming source, and levels of magnesium and potassium in the soil carefully monitored to be sure they are present in reasonable amounts, compared with the high levels of added calcium.

The use of "clean" sludges—those containing low levels of metal and organic contaminants—for agronomic crops is certainly an acceptable practice. Sludges should not be applied to soils when growing crops for direct human consumption unless it can be demonstrated that, in addition to low levels of potentially toxic materials, organisms dangerous to humans are absent.

Application Rates for Organic Materials

The amount of residue added to a soil is often determined by the cropping system. The crop residues can be left on the surface or incorporated by tillage. Different amounts of residue will remain under different crops, rotations, or harvest practices. For example, 3 or more tons per acre of leaf, stalk, and cob residues remain in the field when corn is harvested for grain depending on yield. If the entire plant is harvested to make silage, there is little left except the roots.

When "imported" organic materials are brought to the field, you need to decide how much and when to apply them. In general, application rates of these residues will be based on their probable contribution to the nitrogen nutrition of plants. We don't want to apply too much available nitrogen because it will be wasted. Nitrate from excessive applications of organic sources of fertility may leach into groundwater just as easily as nitrate originating from purchased synthetic fertilizers. In addition, excess nitrate in plants may cause health problems for humans and animals.

Sometimes the fertility contribution of phosphorus may be the main factor governing application rates of organic material. Excess phosphorus entering lakes can cause an increase in the growth of algae and other aquatic weeds, decreasing water quality for drinking and recreation. In locations where this occurs, farmers must be careful to avoid loading the soil with too much phosphorus, from either commercial fertilizers or organic sources.

Effects of Residue and Manure Accumulations

When any organic material is added to soil, it decomposes relatively rapidly at first. Later, when only resistant parts (for example, straw stems high in lignin) are left, the rate of decomposition decreases greatly. This means that although nutrient availability diminishes each year after adding a residue to the soil, there are still long-term benefits from adding organic materials. This can be expressed by using a "decay series." For example, 50, 15, 5, and 2% of the amount of nitrogen added in manure may be released in the first, second, third, and fourth years following addition to soils. In other words, crops in a regularly manured field get some nitrogen from manure that was applied in past years. So, if you are starting to manure a field, somewhat more manure will be needed in the first year than will be needed in years 2, 3, and 4 to supply the same total amount of nitrogen to a crop. After some years, you may need only half of the amount used to supply all the nitrogen needs in the first year. However, it is not uncommon to find farmers who are trying to build up high levels of organic matter actually overloading their soils with nutrients, with potential negative effects on crop quality and the environment. Instead of reducing the amount of off-farm residue with time, they use a standard amount annually. This may lead to excess amounts of nitrate, lessening the quality of many plants and harming groundwater, as well as excess amounts of phosphorus, a potential surface water pollution problem.

Organic Matter Management on Different Types of Farms

Animal-based Farms

It is certainly easier to maintain soil organic matter in animal-based agricultural systems. Manure is a valuable by-product of having animals. Animals also can use sod-type grasses and legumes as pasture, hay, and haylage (hay stored under airtight conditions so that some fermentation occurs). It is easier to justify putting land into perennial forage crops for part of a rotation when there is an economic use for the crops. Animals need not be on the farm to have positive effects on soil fertility. A farmer may grow hay to sell to a neighbor and trade for some animal manure from the neighbor's farm, for example. Occasionally, formal agreements between dairy farmers and vegetable growers lead to cooperation on crop rotations and manure application.

Systems without Animals

It is more challenging, although not impossible, to maintain or increase soil organic matter on non-livestock farms. It can be done by using reduced tillage, intensive use of cover crops, intercropping, living mulches, rotations that include crops with high amounts of residue left after harvest, and attention to other erosion-control practices. Organic residues, such as leaves or clean sewage sludges, can sometimes be obtained from nearby cities and towns. Straw or grass clippings used as mulch also add organic matter when they later become incorporated into the soil by plowing or by the activity of soil organisms. Some vegetable farmers use a "mow-and blow" system in which crops are grown on strips for the purpose of chopping them and spraying the residues onto an adjacent strip. When you use off-farm organic materials such as composts and manures, soil should be tested regularly to ensure that it does not become overloaded with nutrients.

Chapter 9

Soil Erosion

The displacement of the topmost layer of the soil is termed as soil erosion. It is caused due to the activity of water, glaciers, wind, living organisms, etc. The effect of soil erosion on vegetation and slope stability as well as the environment and the ways in which erosion can be managed have been dealt with in this chapter.

Soil erosion is, at its core, a natural process. Put simply, it is when topsoil, which is the upper-most layer of the ground, is moved from one spot to another. Why this matters is because topsoil is the part of the land that is highest in organic matter and best suited for farming and other fertile activities, which is why soil erosion can have the greatest impact on farmers and agricultural land. In other words, soil erosion is a naturally occurring and slow process that refers to loss of field's top soil by water and wind or through conversion of natural vegetation to agricultural land.

When farming activities are carried out, the top soil is exposed and is often blown away by wind or washed away by rain. When soil erosion occurs, the movement of the detached topsoil is typically facilitated by either a natural process – such as wind or water movement – or by the impact of man, such as through tilling farmland.

The process of soil erosion is made up of three parts:

- Detachment: This is when the topsoil is actually "detached" from the rest of the ground.

- Movement: This is when the topsoil is relocated to another area.

- Deposition: Where the topsoil ends up after this process.

When it comes to our planet, natural resources are typically affected by two things – either naturally-occurring ones such as weather, or from man-made influence. Soil erosion, or the gradual reduction of topsoil in a geographic area, can be caused by both natural and unnatural processes, but it can also have great effects on inhabitants of an affected area. One of the major concerns regarding soil erosion is that it can permanently affect the land, which can be devastating for farmers or those with agricultural pursuits.

Unfortunately, many people are still uneducated about soil erosion, which is leading to the occurrence in greater amounts around the world. Soil erosion contributes to pollution in adjacent water sources and reduces cropland productivity. Major crops that cause soil erosion include coffee, cotton, tea, tobacco, palm oil, soybean and wheat that can increase soil erosion beyond the soil's ability to maintain itself.

Causes of Soil Erosion

As mentioned, the predominant causes of soil erosion are either related to naturally-occurring events or influenced by the presence of human activity. Some of the principal causes of soil erosion include:

- Rain and rainwater runoff: In a particular heavy rain, soil erosion is common. First of all, the water starts to break down the soil, dispersing the materials it is made of. Typically, rainwater runoff will impact lighter materials like silt, organic matter, and finer sand particles, but in heavy rainfall, this can also include the larger material components as well.

- Farming: When land is worked through crops or other agricultural processes, it reduces the overall structure of the soil, in addition to reducing the levels of organic matter, making it more susceptible to the effects of rain and water. Tilling in particular, because it often breaks up and softens the structure of soil, can be a major contributor to erosion. Farming practices that reduce this activity tend to have far less issues with soil erosion.

- Slope of the land: The physical characteristics of the land can also contribute to soil erosion. For example, land with a high hill slope will perpetuate the process of rainwater or runoff saturation in the area, particularly due to the faster movement of the water down a slope.

- Lack of vegetation: Plants and crops help maintain the structure of soils, reducing the amount of soil erosion. Areas with less naturally-occurring flora may be a hint that the soil is prone to erosion.

- Wind: Wind can be a major factor in reducing soil quality and promotion erosion, particularly if the soil's structure has already been loosened up. However, lighter winds will typically not cause too much damage, if any. The most susceptible soil to this type of erosion is sandy or lighter soil that can easily be transported through the air.

Effects of Soil Erosion

A major problem with soil erosion is that there is no telling how quickly or slowly it will occur. If largely impacted by ongoing weather or climate events, it may be a slow-developing process that is never even noticed. However, a severe weather occurrence or other experience can contribute to rapid-moving erosion, which can cause great harm to the area and its inhabitants.

Some of the Greatest Effects of Soil Erosion Include

- Loss of topsoil: Obviously, this is the biggest effect of soil erosion. Because topsoil is so fertile, if it is removed, this can cause serious harm to farmer's crops or the ability to effectively work their land.

- Soil compaction: When soil under the topsoil becomes compacted and stiff, it reduces the ability for water to infiltrate these deeper levels, keeping runoff at greater levels, which increases the risk of more serious erosion.

- Reduced organic and fertile matter: As mentioned, removing topsoil that is heavy with organic matter will reduce the ability for the land to regenerate new flora or crops. When new crops or plants can't be placed successfully in the area, this perpetuates a cycle of reduced levels of organic nutrients.

- Poor drainage: Sometimes too much compaction with sand can lead to an effective crust that seals in the surface layer, making it even harder for water to pass through to deeper layers. In some ways, this can help erosion because of the densely packed soil, but if it perpetuates greater levels of runoff from rainwater or flooding, it can negatively impact the crucial topsoil.

- Issues with plant reproduction: When soil is eroded in an active cropland, wind in particular makes lighter soil properties such as new seeds and seedlings to be buried or destroyed. This, in turn, impacts future crop production.

- Soil acidity levels: When the structure of the soil becomes compromised, and organic matter is greatly reduced, there is a higher chance of increased soil acidity, which will significantly impact the ability for plants and crops to grow.

- Long term erosion: Unfortunately, if an area is prone to erosion or has a history of it, it becomes even harder to protect it in the future. The process has already reduced the soil structure and organic matter of the area, meaning that it will be harder to recover in the long run.

- Water pollution: A major problem with runoff from soils – particularly those used for agricultural processes – is that there is a greater likelihood that sediment and contamination like the use of fertilizer or pesticide. This can have significant damage on fish and water quality.

Solutions for Soil Erosion

When it comes to finding solutions for soil erosion, the most useful techniques found tend to be those that emphasize reinforcing the structure of the soil, and reducing processes that affect it.

- Careful tilling: Because tilling activity breaks up the structure of soil, doing less tilling with fewer passes will preserve more of the crucial topsoil.

- Crop rotation: Plenty of crop rotation is crucial for keeping land happy and healthy. This allows organic matter to build up, making future plantings more fertile.

- Increased structure for plants: Introducing terraces or other means of stabilizing plant life or even the soil around them can help reduce the chance that the soil loosens and erodes. Boosting areas that are prone to erosion with sturdy plant life can be a great way to stave off future effects.

- Water control: For those areas where soil erosion is predominantly caused by water – whether natural or man-made – specialized chutes and runoff pipes can help to direct these water sources away from the susceptible areas, helping stave off excess erosion. Having these filters in particular areas rather than leading to natural bodies of water is a focus to reduce pollution.

- Increased knowledge: A major factor for preventing soil erosion is educating more and more people who work with the land on why it is a concern, and what they can do to help reduce it. This means outreach to farmers in susceptible areas for ways that they can help protect crops from inclement weather, or ways that they can help make sure their soil remains compact without restricting their plant growing activities.

Aeolian Processes

Eolian landforms are found in regions of the Earth where erosion and deposition by wind are the dominant geomorphic forces shaping the face of the landscape. Regions influenced by wind include most of the dry climates of the Earth (Figure). According to the Köppen Climate Classification System, this would include regions of the world that are classified as arid deserts (BW) and semiarid steppe (BS). Wind can also cause erosion and deposition in environments where sediments have been recently deposited or disturbed. Such environments include lake and ocean coastline beaches, alluvial fans, and farmland where topsoil has been disturbed by cultivation.

Figure Global distribution of major deposits of eolian derived sediments.

Unlike streams, wind has the ability to transport sediment up-slope as well as down-slope. The relative ability of wind to erode materials is slight when compared to the other major erosional mediums, water and ice. Ice and water can have greater erosive power primarily because of their greater density. Water is about 800 times more dense than air (density of air is 1.29 kg m-3, while the density of water is 1000 kg m-3). This physical difference limits the size of particles wind can move. The power of wind to erode surface particles is controlled primarily by two factors: wind velocity and surface roughness. Erosive force increases exponentially with increases in wind velocity. For example, a velocity increase from 2 to 4 meters per second causes an eight-fold increase in erosive capacity, while an increase in wind speed from 2 to 10 meters per second generates a 125-fold increase in erosional force. Consequently, fast winds are capable of causing much more erosion than slow winds.

At ground level, the roughness of the surface plays an important role in controlling the nature of wind erosion. Boulders, trees, buildings, shrubs, and even small plants like grass and herbs can increase the frictional roughness of the surface and reduce wind velocity. Vegetation can also reduce the erosional effects of wind by binding soil particles to roots. Thus, as a general rule, the areas that show considerable amounts of wind erosion are open locations with little or no plant cover.

Threshold and Terminal Fall Velocities

Threshold velocity can be defined as velocity required to entrain a particle of a particular size. In general, the larger the particle, the higher the threshold velocity required to move it. This law can sometimes be broken when clay sized particles are involved in the entrapment process. Clay particles have a general tendency to become cohesively bonded to each other. This aggregation results in the clumping of several particles into a mass of much larger size. As a result, the threshold velocity required to entrain clay is as great as the wind speed required to move grains of sand. Silt is usually the easiest type of particle to be entrained by wind.

Terminal fall velocity can be defined as velocity at which a particle being transported by wind or water falls out and is deposited on the ground surface. Figure describes the terminal fall velocities for clay, silt, and sand sized particles for wind. On this semi-log graph, a simple, somewhat linear, relationship is observed. The larger the particle the greater the wind speed that is required to keep it moving above the ground surface.

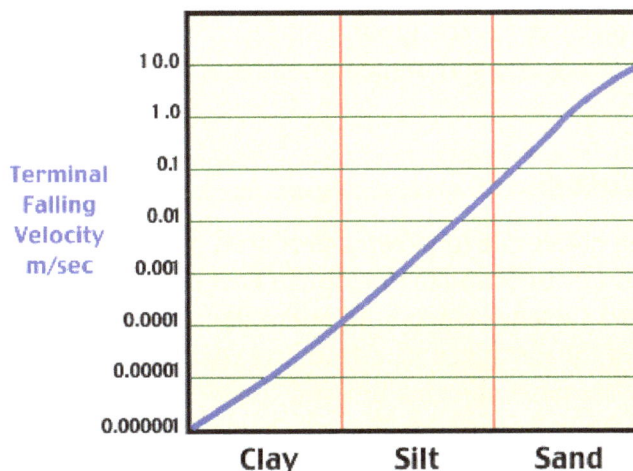

Figure Falling velocities for clay, silt, and sand sized particles for wind. Note the fall velocity for clay is many orders of magnitude less than the fall velocity for sand.

Sand Transport

Most of our present knowledge about the wind's ability to erode and transportation sand comes from the 1920s to 40s work of Ralph Bagnold in the deserts of North Africa. In his numerous observations and experiments dealing with sand movement, Bagnold discovered many of the key principles controlling the erosion and transport of sand in deserts.

Three different processes are responsible for the transport of sediment by wind. Wind erosion of surface particles begins when air velocities reach about 4.5 meters per second. A rolling motion called traction or creep characterizes this first movement of particles. In strong winds, particles as large as small pebbles can move through traction. About 20 to 25 percent of wind erosion is by traction. The second type of wind sediment transport involves particles being lifted off the ground, becoming suspended in the air, and then returning to the ground surface several centimeters downwind. This type of transport is called saltation, and this process accounts for 75 to 80 percent of the sediment transport in dry land environments. Saltating particles are also responsible for sending additional sediment into transport. When a falling particle strikes the ground surface, part of its force of impact is transferred to another particle causing it to become airborne. Small sized particles like silt and clay have the ability to be lifted well above the zone of saltation during very strong winds and can be carried in suspension thousands of meters into the air and hundreds of kilometers downwind.

Erosional Landforms

When the force of wind is concentrated on a particular spot in the landscape, erosion can carve out a pit known as a deflation hollow. Deflation hollows range in size from a few meters to a hundred meters in diameter, and may develop over several days or a couple of seasons. Much larger depressions are also found in the arid regions throughout the world. These broad, shallow depressions, called pans, can cover thousands of square kilometers. One of the largest pans, known as the Qattara Depression is found in the Lybian Desert of Egypt. This pan covers around 15,000 square kilometers.

In some dry climate areas, persistent winds erode all sediments the size of sand and smaller leaving pebbles and larger particles on the ground surface. Surfaces loaded with such particles are called desert pavement or reg and sometimes resemble a worn, polished cobblestone street surface.

Depositional Landforms

Sand dunes are the most noticeable landforms produced by wind erosion and deposition of sediment. The largest dune fields are found in the Middle East and North Africa. Most large dune fields act more or less as closed systems. Once sand enters these systems, it does not leave. However, dune fields do shift across the Earth's surface from time to time. Periodic migrations of dune fields are normally caused by seasonal changes in wind direction. Over longer periods of time, dune fields may expand or contract because of climatic change. In the last few decades scientists have noticed a spatial expansion of deserts that may be correlated to human disturbance of natural vegetation cover because of agriculture.

Sand Dune Formation

Sand dunes form in environments that favor the deposition of sand. Deposition occurs in areas where a pocket of slower moving air forms next to much faster moving air. Such pockets typically form behind obstacles like the leeward sides of slopes. As the fast air slides over the calm zone, saltating grains fall out of the air stream and accumulate on the ground surface.

Figure: Wind ripples on top of much larger sand dunes. Wind ripples are mini-dunes between 5 centimeters and 2 meters apart and 0.1 to 5 centimeters in height.

They are created by saltation when the sand grains are of similar size and wind speed is consistent. A series of wind ripples is initiate by a single irregularity in the ground surface. This irregularity launches the grains in the air and the consistencies of size and wind speed cause saltation at repeated regular intervals downwind.

Dunes first begin their life as a stationary pile of sand that forms behind some type of vertical obstacle. However, when they reach a certain size threshold continued growth may also be associated with active surface migration. In a migrating dune, grains of sand are transported by wind from the windward to the leeward side and begin accumulating just over the crest. When the upper leeward slope reaches an angle of about 30 to 34 degrees the accumulating pile becomes unstable, and small avalanches begin to occur, moving sand to the lower part of the leeward slope. As a result of this process, the dune migrates over the ground as sand is eroded from one side and deposited on the other. This process also causes the appearance of the dune to take on a wave shape. Active movement of sand particles across the dune causes windward slope to become shallow, while the leeward slope maintains a steep slope or slip-face.

The velocity of the wind above the ground surface determines the height of a dune. The maximum height is variable but usually falls in the range of 10 to 25 meters. In most cases, dune height is a function of surface friction. Height growth stops when friction can no longer slow the wind flowing over the dune to a point where deposition occurs. The tallest sand dunes in the world are found in Saudi Arabia and measure more than 200 meters. However, these features are not individual dunes, but a massive complex of sand dunes that forms when smaller, faster moving dunes migrate onto larger, slower moving dunes.

Desert Dunes

Desert sand dunes occur in an amazing diversity of forms. Table describes the major types of dunes classified by geomorphologists.

Table: Major types of sand dunes.

Type	Description
Barchan	Crescent-shaped dune whose long axis is transverse to the dominant wind direction. The points of the dune, called the wings of the barchan, are curved downwind and partially enclosing the slip-face. Barchans usually form where there is a limited supply of sand, reasonably flat ground, and a fairly even flow of wind from one direction. Single slip-face.
Transverse	Long asymmetrical dunes that form at right angles to the wind direction. Form when there is an abundant supply of sand and relatively weak winds. These dunes have a single long slip-face.
Parabolic	Crescent-shaped dune whose long axis is transverse to the dominant wind direction. The points of this dune curve upwind. Multiple slip-faces. These dunes form when scattered vegetation stabilizes sediments and a U-shaped blowout forms between clumps of plants.
Barchanoid Ridge	Is a long, asymmetrical dune that runs at right angles to the prevailing wind direction. A barchanoid ridge consists of several joined barchan dunes and looks like a row of connected crescents. Each of the barchan dunes produces a wave in the barchanoid ridge. Occurs when sand supply is greater than in the conditions that create a barchan dune.
Longitudinal	Sinuous dune that can be more than 100 kilometers long and 100 meters high. Created when there are strong winds from at least two directions. The dune ridge is symmetrical, aligned parallel to the net direction of the wind, and has slip-faces on either side.
Seif	Sub-type of longitudinal dune that is shorter and has a more sinuous ridge.
Star Dune	Large pyramidal or star-shaped dune with three or more sinuous radiating ridges from a central peak of sand. This dune has 3 or more slip-faces. Produced by variable winds. This dune does not migrate along the ground, but does grow vertically.
Dome	Mound of sand that is circular or elliptical in shape. Have no slip-faces. May be formed by the modification of stationary barchans.
Reversing	Dune that is intermediate between a star and transverse dunes. Ridge is asymmetrical and has two slip-faces.

Figure: Longitudinal dunes, Arabian Peninsula.

Coastal Dunes

Active sand dune formation is also found on the coasts of the continents. Coastal dunes form when there is a large supply of beach sand and strong winds blowing from sea to shore. The beach area must also be wide and sufficiently influenced by wave action to keep it free of plants.

Many coastal dune deposits develop in association with blowouts in ridges of beach sand. Blowouts are small saucer shaped depressions where there is a deposit of sand at the upwind end of the feature. As wind erosion continues, the deposit grows and begins to migrate inland forming a parabolic sand dune. The flanks of these dunes tend to be more stable and are often colonized by plants like dune grass, sea oats, and sand cherry. This colonization by plants re-inforces the stability of the dune's flanks.

Coastal dunes are dissimilar from desert dunes in their form and shape and the fact that they do not migrate. The presence of vegetation limits migration and radically modifies the dune environment by altering the patterns of airflow, reducing sand erosion, and stabilizing the lee slope of the dune.

Figure: Vegetated beach dune.

Loess Deposits

Loess is another major deposit created by wind. Less visible than sand dunes, loess is found over large areas of the Earth. It is also important for humans because it creates very fertile soils. Large deposits of loess exist in northeastern China, central plains of the United States, Pampas of Argentina, the Ukraine, and central Europe. Loess is mainly composed of silt. Because of its small size it can be held in suspension and carried great distances by wind. Most loess deposits appear to have been formed by winds that blew over glacial deposits during the Pleistocene. The major deserts of the world also appear to have produced significant amounts of loess. Recent research has uncovered that soils in the Amazon basin may have been enriched with loess deposits that originated from African deserts.

Vegetation and Slope Stability

Vegetation (trees, shrubs, grasses, flower and ground covers) helps stabilize slopes in numerous ways. Native plants provide wildlife habitat, are adapted to our native soils, can adapt better to climate disruptions, and interact with each other in ways hybrid/exotic plant communities do not.

Vegetation management of urban, forested, coastal, marine shorelines, freshwater shorelines, and other riparian zones should conserve and maintain plant cover to reduce surface soil erosion and landslide styled movements of soil. The relative effectiveness of vegetation in any site will be a function of quality of vegetation, topography, slope, hydrology, geology and soils. Surface Soil Erosion Ways in which vegetation and specifically herbaceous plants prevent soil erosion include:

- Interception – foliage and plant residues absorb rainfall energy and prevent soil compaction from raindrops.

- Restraint – root system physically binds or restrains soil particles while above ground residues filter sediment out of rain and storm water runoff.

- Retardation – above-ground residues increase surface roughness and slow velocity of runoff.

- Infiltration – roots and plant residues help maintain soil porosity and permeability.

- Transpiration – depletion of soil moisture by plants delays onset of saturation and runoff.

Erosion occurs when rainfall dislodges soil particles and carries them off a slope, forming rills and gullies that can trigger landslides. Raindrops hitting the soil surface can also seal the soil particles and make a crust that prevents infiltration and increases surface flow speeds, creating runoff. Of all these processes, retardation (slowing) of mobile surface waters and infiltration (increased soil porosity) both substantially reduces impacts to particle soil erosion. Soil structure also plays a substantial role. Interception (absorption) of rainfall energy and transpiration (soil moisture wicking) reduce heavily in the winter when plants drop their leaves. Evergreen plants, as opposed to deciduous plants, do retain interception qualities throughout the winter, but their ability to wick moisture from the ground is still heavily reduced.

The primary factors of vegetation that affect mass-movements in slopes, particularly shallow sliding in slopes include:

- Root reinforcement – roots mechanically reinforce a soil transfer of shear stresses in the soil to tensile resistance in the roots.

- Soil moisture modification – Transpiration and interception by the foliage limit buildup of soil moisture stress.

- Buttressing and arching – anchored and embedded stems can act as buttress piles or arch abutments in a slope, counteracting shear stresses.

- Wind-throwing – destabilizing influence from turning moments exerted on a slope as a result of strong winds blowing downslope through trees.

Root reinforcement, soil moisture modification (reduction), and buttressing and arching normally increase slope stability. Surcharge and wind-throwing are having a net destabilizing effect.

Erosion Control

Erosion is the loss of soil. As soil erodes, it loses nutrients, clogs rivers with dirt, and eventually turns the area into a desert. Although erosion happens naturally, human activities can make it much worse. Erosion can turn once healthy, vibrant land into arid, lifeless terrain and further cause landslides and mudslides. Erosion can be controlled easily on a construction site when the right means, tools, and methods are used at the right time. The most natural and effective way to prevent erosion control is by planting vegetation. Roots from plants, especially trees, grip soil and will effectively prevent the excess movement of soil throughout the ground. Another popular erosion control method is the use of a silt fence. A silt fence is a long fabric barrier that is installed along a hill, and collects any storm water that would carry loose soil Another effective technique used for soil erosion control is erosion control matting. Erosion control matting is laid on top of loose soil and is secured into place.

Erosion Control Methods

This illustration shows the major methods employed in controlling soil erosion with reference to runoff factors.

There are two basic approaches as:

1. Reducing runoff amount and

2. Reducing runoff velocity.

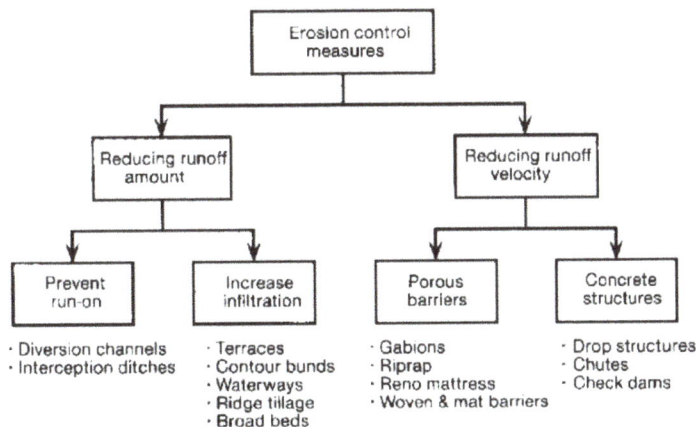

Five Main Techniques

There are five main techniques that can be used in controlling soil erosion are.

They are as follows:

(i) Contour bunding and Farming

(ii) Strip Cropping

(iii) Terracing

(iv) Gully Reclamation

(v) Shelter Belts.

Soil erosion can be controlled by adopting land management practices and also by changing the pattern of some human activities which accelerate soil erosion. One such idea is to minimize disturbance.

Land Disturbing Activities

The most effective form of erosion control is to minimize the area of disturbance. The land disturbing activities are the following:

a) Quarries:

Quarries are places of naturally occurring hard rock that is mined for rock and gravels.

The products from quarry operations are used for roading, building and in rock. Protections measures, i.e., rip-rap. The following specific issues associated with quarry operations:

- Road access

- Storm water

- Overburden disposal

- Stockpile areas

- Rehabilitation of worked out areas

- Riparian protection areas

- Maintenance schedule for erosion and sediment control treatment structures.

b) Trenching:

Trenching (usually for installing utility services), often occurs at the end of bulk earthworks.

Topsoil and sub-soils should be stockpiled separately adjacent to the trench so that at the completion of the operation these soils can be replaced in the appropriate order and vegetation established.

c) Clean fills:

Clean fills dispose of unwanted fill material which may contain other material.

d) Roading:

The linear nature of roading poses challenges for erosion and sediment control measures. They need to be planned to ensure controls are successful.

Minimize Disturbance

The most effective form of erosion control is to minimize the area of disturbance, retaining as much existing vegetation as possible. This is especially important on steep slopes or in the vicinity of water bodies, where no single measure will adequately control erosion and where receiving environments may be highly sensitive. Match land development to land sensitivity. Watch out for and avoid areas that are wet (streams, wetlands, springs), have steep or fragile soils. Analyze all the "limits of disturbance".

a) Stage Construction:

Temporary stockpiles, access and utility service installation all need to be considered.

b) Protect Steep Slopes:

Steep slopes should be avoided where practicable.

c) Protect Water bodies:

All water bodies and proposed drainage patterns.

Map all water bodies and show limits of disturbance and protection measures.

d) Stabilize Exposed Areas Rapidly:

Conventional sowing to mulching. Mulching is an effective instant protection.

e) Install Perimeter Controls:

Perimeter controls above the site keep clean water runoff out of the worked area. Common controls are diversion drains, silt fences and earth bunds.

f) Employ Detention Devices:

Earthworks will still discharge sediment-laden runoff during storms.

g) Runoff Diversion Channel/Bund:

This is a non-erodible channel or bund constructed for the conveyance of runoff constructed to a site specific cross section and grade design. It is done to either protect work areas from upslope runoff, or to divert sediment laden water to an appropriate sediment retention structure.

h) Contour Drain:

It is a temporary ridge or excavated channel, or combination of ridge and channel, constructed to convey water across sloping land on a minimal gradient. To periodically break overland flow across disturbed areas in order to limit slope length and thus the erosive power of runoff and to divert sediment laden water to appropriate controls or stable outlets.

i) Rock Check Dam:

Small temporary dam constructed across a channel (excluding perennial water bodies), usually in series, to reduce flow velocity. It may also retain coarse sediment. Check dams are constructed in order to reduce the velocity of concentrated flows, thereby reducing erosion of the channel. Rock check dams will trap some sediment, but they are not designed as a sediment retention measure.

j) Level Spreader:

A non-erosive outlet to disperse concentrated runoff uniformly across a slope. The level spreader provides a relatively low cost option, which can convert concentrated flow to sheet flow and release it uniformly over a stabilized area.

k) Pipe Drop Structure/Flume:

A temporary pipe structure or constructed flume placed from the top of a slope to the bottom of a slope. A pipe drop structure or a flume structure is installed to convey surface runoff down the face of unsterilized slopes in order to minimize erosion on the slope face.

l) Benched Slope:

Modification of a slope by reverse sloping to divert runoff to an appropriate conveyance system. To limit the velocity and volume and hence the erosive power of water flowing down a slope and therefore minimizing erosion of the slope face.

m) Surface Roughening:

Roughening a bare earth surface with horizontal grooves running across a slope or tracking with construction equipment. To aid in the establishment of vegetative cover from seed, to reduce runoff velocity, to increase infiltration, to reduce erosion and assist in sediment trapping.

n) Stabilized Construction Entrance:

A stabilized pad of aggregate on a filter cloth base located at any point where traffic will be entering or leaving a construction site. To prevent site access points from becoming sediment sources and to assist in minimizing dust generation and disturbance of areas adjacent to the road frontage by giving a defined entry/exit point.

o) Geosynthetic Erosion Control Systems (GECS):

The protection of channels and erodible slopes utilizing artificial erosion control material such as geosynthetic matting, geotextiles or erosion matting. To immediately reduce the erosion potential of disturbed areas and/or to reduce or eliminate erosion on critical sites during the period necessary to establish protective vegetation. There are both Temporary and Permanent Non-Degradable GECS.

Revegetation Techniques

a) Top Soiling:

The placement of topsoil over prepared subsoil prior to the establishment of vegetation. To provide a suitable soil medium for vegetative growth while providing some limited short term erosion control capability.

b) Temporary and Permanent Seeding:

The planting and establishment of quick growing and/or perennial vegetation to provide temporary and/or permanent stabilization on exposed areas. Temporary seeding is designed to stabilize the soil and to protect disturbed areas until permanent vegetation or other erosion control measures can be established.

c) Hydroseeding:

Hydroseeding is a planting process that uses slurry of seed and mulch. It is often used as an erosion control technique. The application of seed, fertilizer and a paper or wood pulp with water in the form of slurry which is sprayed over the area to be revegetated. To establish vegetation quickly while providing a degree of instant protection from rain drop impact.

d) Mulching:

Mulches are loose coverings or sheets of material placed on the surface of cultivated soil. Organic mulches also improve the condition of the soil. As these mulches slowly decompose, they provide organic matter which helps keep the soil loose. This improves root growth, increases the infiltration of water, and also improves the water-holding capacity of the soil. The application of a protective layer of straw or other suitable material to the soil surface. To protect the soil surface from the erosive forces of raindrop impact and overland flow. Mulching assists in soil moisture conservation, reduces runoff and erosion, controls weeds, prevents soil crusting and promotes the establishment of desirable vegetation.

e) Turfing:

A surface layer of earth containing a dense growth of grass and its matted roots; sod. Turfing is an artificial substitute for such a grassy layer, as on a playing field. The establishment and permanent stabilization of disturbed areas by laying a continuous cover of grass turf. To provide immediate vegetative cover to stabilize soil on disturbed areas.

Sediment Control Measures

Sediment Retention Pond

A temporary pond formed by excavation into natural ground or by construction of an embankment and incorporating a device to dewater the pond at a rate that will allow suspended sediment to settle out. To treat sediment-laden runoff and reduce the volume of sediment leaving a site, thus protecting downstream environments from excessive sedimentation and water quality degradation.

Chemical Flocculation Systems

A treatment system designed to add a flocculating chemical to sediment retention ponds.

Used to increase the sediment capture performance of sediment retention ponds by causing suspended sediment to "clump" resulting in faster settling rates.

Silt Fence

The purpose of a silt fence is to retain the soil on disturbed land. The three principal aspects of silt fence design are: proper placement of fencing, adequate amount of fencing, and appropriate materials. A silt fence is a temporary sediment barrier made of porous fabric. It's held up by wooden or metal posts driven into the ground, so it's inexpensive and relatively easy to remove. The fabric ponds sediment-laden storm water runoff, causing sediment to be retained by the settling processes. A temporary barrier of woven geotextile fabric is also used to intercept sediment laden runoff from small areas of soil disturbance.

Super Silt Fence

A temporary barrier of woven geotextile fabric over chain link fence used to intercept sediment laden runoff from soil disturbance in small catchment areas. A super silt fence provides more robust sediment control compared with a standard silt fence and allows up to four times the catchment area to betreated by an equivalent length of standard silt fence.

Stormwater Inlet Protection

A barrier across or around a cesspit (stormwater inlet). To intercept and filter sediment-laden runoff before it enters a reticulated stormwater system via a cesspit, thereby preventing sediment-laden flows from entering receiving environments.

Decanting Earth Bund

A temporary berm or ridge of compacted earth constructed to create impoundment areas where ponding of runoff can occur and suspended material can settle before runoff is discharged. Used to intercept sediment-laden runoff and reduce the amount of sediment leaving the site by detaining sediment-laden runoff.

Decanting Topsoil Bund

A temporary berm or ridge of track rolled topsoil, constructed to create impoundment areas where ponding of runoff can occur and suspended material can settle before runoff is discharged. Used to intercept sediment-laden runoff from small areas (less than 0.3 ha) and reduce the amount of sediment leaving the site by detaining sediment-laden runoff.

Sump/Sediment Pit

A temporary pit which is constructed to trap and filter water before it is pumped to a suitable discharge area. To treat sediment-laden water that has been removed from areas of excavation or areas where ponded sediment-laden water cannot drain by other means.

Riprap

Rock pieces are piled up to create a structure called as rip-rap. These are rubble composed of a variety of rock types including limestone and granite, which are used to armor embankments, shorelines, bridge abutments, streambeds and other seaside constructions to prevent soil erosion due to concentrated runoff or other water-related causes. A limitation of riprap arises when the slopes of the considered area are greater than 2:1; the rubble becomes unstable and is itself prone to erosion. In these circumstances, gabions are used.

Gabions

Gabion is an Italian word gabbia meaning "cage". The gabions are riprap encased in galvanized, steel-wire mesh cages or cylinders. These are used to stabilize slopes, stream banks, or shorelines against erosion. They are usually placed on slopes at an angle—either battered or stepped back, rather than stacked vertically. The life expectancy of gabions rely entirely on their wire frames, and premium ones have a guaranteed structural consistency of fifty years.

Buffer Strip

These are narrow areas of land maintained in permanent vegetation to trap sediment, slow down runoff, and even control air, soil, and water quality. The root systems of the vegetation anchor soil particles together which help stop the soil from being eroded by winds. They also reduce the risk of landslides and other slower forms of erosion by stabilizing stream banks.

Soil Binders

Soil binders bind soil particles together in order to make the soil matrix more water and pressure resistant. Soil binder has two functions: erosion control and soil stabilization. The success of common soil binder applications varies significantly depending on the local conditions and use of stabilized soil. Soil binders have multiple purposes: soil stabilization, dust control and erosion control. Some soil binder products can combat all these issues at the same time. Cement is commercial soil binder although it has numerous drawbacks. Lime soil binder products are quicklime, hydrated lime and lime slurry. Fly ash is typically used to stabilize sub base or subgrade, and is not among soil binder products suitable for surfacing due to low resistance to abrasive action of traffic. Fly ash application has adverse effect on environment.

Works in Water Bodies

Works within water bodies have a high potential for erosion and discharge of sediment. This is because work is undertaken in or near flowing water - the major cause of erosion. Flowing water causes ongoing scour and provides the transport mechanism to allow sediment to be dispersed downstream of works.

a) Temporary Water body Diversions:

A short term water body diversion to allow works to occur within the main channel under dry conditions. To enable water body diversion without working in wet conditions and without allowing sediment discharges into a water body.

b) Temporary Water body Crossings:

A bridge, culvert or ford installed across a water body for short-term use. To provide a means to cross water bodies without moving sediment into the water body, damaging the bed or channel, or causing flooding during the construction, maintenance or removal of the structure.

Soil Conservation Methods

The pre-eminent methods of soil conservation are:

1. Expansion of vegetative cover and protective afforestation,

2. Controlled grazing,

3. Flood control,

4. Prohibition of shifting cultivation,

5. Proper land utilization,

6. Maintenance of soil fertility,

7. Land reforms, reclamation of wasteland,

8. Establishment of soil research institute and training of soil scientists, and

9. Effective agencies for soil management.

Effect on Environment

Loss of Arable Land

Lands used for crop production have been substantially affected by soil erosion. Soil erosion eats away the top soil which is the fertile layer of the land and also the component that supports the soil's essential microorganisms and organic matter. In this view, soil erosion has severely threatened the productivity of fertile cropping areas as they are continually degraded.

Because of soil erosion, most of the soil characteristics that support agriculture have been lost, causing ecological collapse and mass starvation. It is likely that most of the cultivated areas around the globe are vulnerable to soil erosion.

Water Pollution and Clogging of Waterways

Soils eroded from agricultural lands carry pesticides, heavy metals, and fertilizers which are washed into streams and major water ways. This leads to water pollution and damage to marine and freshwater habitats. Accumulated sediments can also cause clogging of water ways and raises the water level leading to flooding.

The water quality of various streams, rivers, and coastal areas has also been deteriorated as a result of soil erosion, eventually affecting the health of the local communities.

Sedimentation and Threat to Aquatic Systems

Apart from polluting the water systems, high soil sedimentation can be catastrophic to the survival of aquatic life forms. Silt can smother the breeding grounds of fish and equally lessens their food supply since the siltation reduces the biodiversity of algal life and beneficial aquatic plants. Sediments may also enter the fish gills, affecting their respiratory functions.

Air Pollution

Wind erosion picks up dust particles of the soil and throws them into the air, causing air pollution. Some of the dust particles may contain harmful and toxic particles such as petroleum and pesticides that can pose a severe health hazard when inhaled or ingested.

Dust plumes from the deserts or dry areas can cause large and widespread air pollution as the winds move. Such a case is evident in North America where dust winds from the Gobi desert have recurrently been a serious problem.

Destruction of Infrastructure

Soil erosion can affect infrastructural projects such as dams, drainages, and embankments. The accumulation of soil sediments in dams/drainages and along embankments can reduce their operational lifetime and efficiency. Also, the silt up can support plant life that can, in turn, cause cracks and weaken the structures. Soil erosion from surface water runoff often causes serious damage to roads and tracks, especially if stabilizing techniques are not used.

Desertification

Soil erosion is a major driver of desertification. It gradually transforms a habitable land and the ASAL regions into deserts. The transformations are worsened by the destructive use of the land and deforestation that leaves the soil naked and open to erosion. This usually leads to loss of biodiversity, alteration of ecosystems, land degradation, and huge economic losses.

References

- Slope-stability-vegetation, ecosystem-restoration: peninsulaurbanforestry.com, Retrieved 11 June 2018
- Causes-effects-solutions-of-soil-erosion: conserve-energy-future.com, Retrieved 12 March 2018
- Methods-of-Controlling-Soil-Erosion-315785523: researchgate.net, Retrieved 08 July 2018
- Fundamentals: physicalgeography.net, Retrieved 28 April 2018
- Causes-and-effects-of-soil-erosion, environment: eartheclipse.com, Retrieved 25 March 2018

Chapter 10

Soil Pollution

The contamination of the soil caused due to any change to the natural soil environment or by the presence of xenobiotic chemicals in the soil is termed as soil pollution. It is caused due to an improper disposal of waste, industrial activities or agricultural chemicals. This chapter analyzes the causes of soil pollution, its effects and the measures to control soil pollution.

"Soil pollution" refers to the presence of a chemical or substance out of place and/ or present at a higher than normal concentration that has adverse effects on any non-targeted organism. Although the majority of pollutants have anthropogenic origins, some contaminants can occur naturally in soils as components of minerals and can be toxic at high concentrations. Soil pollution often cannot be directly assessed or visually perceived, making it a hidden danger.

The diversity of contaminants is constantly evolving due to agrochemical and industrial developments. This diversity, and the transformation of organic compounds in soils by biological activity into diverse metabolites, makes soil surveys to identify the contaminants both difficult and expensive. The effects of soil contamination also depend on soil properties since these controls the mobility, bioavailability, and residence time of contaminants.

Industrialization, wars, mining and intensification in agriculture have left a legacy of contaminated soils around the world. Since urban expansion, soil has been used as a sink for dumping solid and liquid wastes. It was considered that once buried and out of sight, the contaminants would not pose any risk to human health or the environment and that they would somehow disappear. The main sources of soil pollution are anthropogenic, resulting in the accumulation of contaminants in soils that may reach levels of concern.

Soil pollution is an alarming issue. It has been identified as the third most important threat to soil functions in Europe and Eurasia, fourth in North Africa, fifth in Asia, seventh in the Northwest Pacific, eighth in North America, and ninth in sub-Saharan Africa and Latin America. The presence of certain pollutants may also produce nutrient imbalances and soil acidification, two major issues in many parts of the world, as identified in the Status of the World's Soil Resources Report.

The unique global estimate of soil pollution was done in the 1990s by the International Soil Reference and Information Centre (ISRIC) and the United Nations Environment Programme (UNEP), which estimated that 22 million hectares had been affected by soil pollution. Latest data, however, indicate that this number may underestimate the nature and extent of the problem. National attempts to estimate the extent of soil pollution have been undertaken mainly in developed countries. According to the Chinese Environmental Protection Ministry, 16 percent of all Chinese soils and 19 percent of its agricultural soils are categorized as polluted. There are also approximately 3 million potentially polluted sites in the European Economic Area and cooperating countries in the West Balkans and more than 1 300 polluted or contaminated sites in the United States of America (USA) are included on the Superfund National Priorities List. The total number of contaminated sites is estimated at 80 000 across Australia. While these numbers are informative in helping us understand the effects of certain activities on soils, they do not reflect the complete extent of soil pollution around the world, and they highlight the inadequacy of available information and the differences in registering polluted sites across geographic regions. In low- and middle-income countries, the lack of data and information makes one of the world's biggest global problems invisible to the international community. With this overview, it is evident that there is an urgent need to implement a global assessment of soil pollution.

Fortunately, awareness on the importance of soil pollution is increasing around the world, leading to an increase in research conducted on the assessment and remediation of soil pollution. The Revised World Soil Charter recommends that national governments implement regulations on soil pollution and limit the accumulation of contaminants beyond established levels in order to guarantee human health and well-being. Governments are also urged to facilitate remediation of contaminated soils that exceed levels established to protect the health of humans and the environment. Soil pollution took center stage at the Fifth Global Soil Partnership (GSP) Plenary Assembly. Recently, the United Nations Environmental Assembly (UNEA-3) adopted a resolution calling for accelerated actions and collaboration to address and manage soil pollution in the framework of Sustainable Development. This consensus, achieved by more than 170 countries, is a clear sign of the global relevance of pollution and of the willingness of these countries to develop concrete solutions to address pollution problems. At the national level, many countries around the world have adopted or are currently adopting national regulations to protect their soils, to prevent pollution and to address historic problems of contamination. During the Estonian presidency of the Council of the European Union in the second half of 2017, soil became one of the main topics within European discussions, focusing on the key role soils play in food production. In China, soil pollution concerns have grown over the last few years, partly because the problem is directly related to human health. Other developing countries have also recently adopted regulations to prevent and control soil pollution, and to determine soil quality.

The term "soil contamination" has frequently been used as a synonym for soil pollution. The Intergovernmental Technical Panel on Soils (ITPS) under the Global Soil Partnership (GSP) has formalized definitions of the two terms. Soil contamination occurs when the concentration of a chemical or substance is higher than would occur naturally but is not necessarily causing harm. Soil pollution, on the other hand, refers to the presence of a chemical or substance out of place and/or present at a higher than normal concentration that has adverse effects on any non-targeted organism.

One issue is the difficulty in establishing a definition of "normal concentrations." It can be easier to establish hazardous concentrations for human-made substances that do not naturally occur in soils, but it can be challenging to do the same for heavy metals and metalloids, which can originate from the weathering of rocks and minerals. In that case, the parent material, climate and weathering rate need to be taken into consideration before establishing thresholds. Additionally, land use and management practices can affect background levels of substances occurring in soils. When referring to recommended levels, there are also many differences from country to country and among regions, not only about the value itself, but also about the name used to define it, including screening values, threshold values, acceptable concentrations, target values, intervention values, clean-up values, and many others. For that reason, to carry out a global study on the actual state of soil pollution and to be able to make comparisons is extremely complex. However, this is one of the main challenges when making a regional or global assessment of soil pollution.

Agreement among scientists regarding concepts and definitions would help policy-makers and stakeholders to identify other strategies and techniques used in different parts of the world to assess and to address soil pollution. Using a common and a simplified language would also lead to better understanding of the issue of soil pollution.

Point-Source and Diffuse Soil Pollution

Soil pollution, as has been said, can result from both intended and unintended activities. These activities can include the direct deposition of contaminants into the soil as well as complex environmental processes that can lead to indirect soil contamination through water or atmospheric deposition.

Point-source Pollution

Soil pollution can be caused by a specific event or a series of events within a particular area in which contaminants are released to the soil, and the source and identity of the pollution is easily identified. This type of pollution is known as point-source pollution. Anthropogenic activities represent the main sources of point-source pollution. Examples include former factory sites, inadequate waste and wastewater disposal, uncontrolled landfills, excessive application of agrochemicals, spills of many types, and many others. Activities such as mining and smelting that are carried out using poor environmental standards are also sources of contamination with heavy metals in many regions of the world. Other examples of point-source pollution are aromatic hydrocarbons and toxic metals, which are related to oil products. The sites range from leakage from tank installations in Greenland, which caused aromatic hydrocarbon and toxic metal levels that, exceeded the Danish environmental quality criteria, to accidental leakage from oil refinery storage tanks in Tehran.

Point-source pollution is very common in urban areas. Soils near roads have high levels of heavy metals, polycyclic aromatic hydrocarbons, and other pollutants. Old or illegal landfills, where waste is not disposed of properly or according to its toxicity (e.g. batteries or radioactive waste), as well as disposal of sewage sludge and wastewater, can also be important point-source pollutants. Finally, point-source pollution caused by industrial activities can pose risks to human health. For example, over 5 000 brownfields in China are currently affecting the health of their inhabitants. Urban brownfields, located in urban centers, are sites that once harboured industrial activities that have since been relocated.

Diffuse Pollution

Diffuse pollution is pollution that is spread over very wide areas, accumulates in soil, and does not have a single or easily identified source. Diffuse pollution occurs where emission, transformation and dilution of contaminants in other media have occurred prior to their transfer to soil. Diffuse pollution involves the transport of pollutants via air-soil-water systems. Complex analyses involving these three compartments is therefore needed in order adequately to assess this type of pollution. For that reason, diffuse pollution is difficult to analyze, and it can be challenging to track and to delimit its spatial extent. Many of the contaminants that cause local pollution may be involved in diffuse pollution, since their fate in the environment is not well understood. Examples of diffuse pollution are numerous and can include sources from nuclear power and weapons activities; uncontrolled waste disposal and contaminated effluents released in and near catchments; land application of sewage sludge; the agricultural use of pesticides and fertilizers which also add heavy metals, persistent organic pollutants, excess nutrients and agrochemicals that are transported downstream by surface runoff; flood events; atmospheric transport and deposition; and/or soil erosion. Diffuse pollution has a significant impact on the environment and human health, although its severity and extent are generally unknown.

It has been widely demonstrated that the upper layers of soil are enriched in many metals and other elements that are linked to atmospheric deposition from natural and anthropogenic sources. Almost every soil of the northern hemisphere contains radionuclides in higher concentrations than the background level, even in remote areas of North America and Eastern Asia. Due to the nuclear fallout after the catastrophic Chernobyl accident, radionuclides will be present in soils for centuries. More than 50 years will be needed to reach a reduction of 50 percent of the radionuclides, such as Pu or Am, in areas up to 200 km away from Chernobyl.

Due to these different types of pollution from diverse sources, an increase in scientific and technical efforts is needed to develop new methods for measuring, monitoring and better understanding atmospheric deposition processes and the extent of diffuse pollution.

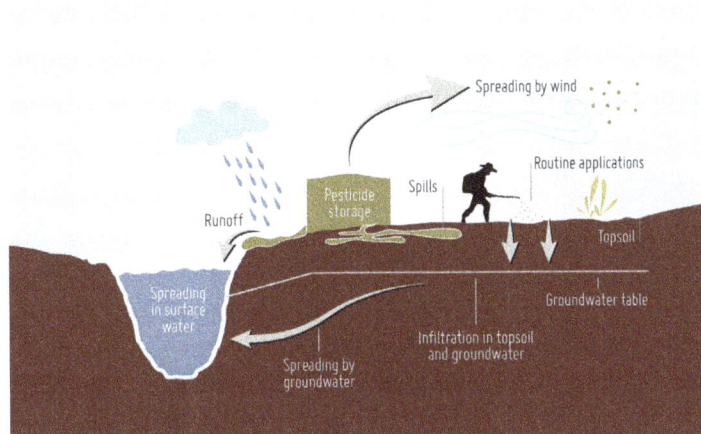

Transport pathway of pesticides in the environment.

Causes of Soil Pollution

Industrial Wastes

The disposal of industrial solid wastes is the major source of soil pollution by toxic chemicals.

The industrial wastes are mainly discharged from coal and mineral mining industries, metal processing industries and engineering industries. They contain toxic metals such as lead, copper and chemicals having acids and are responsible for soil pollution.

It has been reported that about 50 per cent of raw materials ultimately become waste products in industry and about 15 per cent of it are toxic. The chemicals discharged from the industries often enter the surface or groundwater or poison the soil or crops. The production of consumer goods also involved environmental problems, including land and soil pollution.

The following table lists some of the effects of the production of consumer goods:

Table: Production of Consumer Goods and its Effects on Land/Soil

Process	Activity	Effects
Extraction	Mining and oil drilling	Surface land disruption, acid mine drainage, mine failings, sludge ponds, oil spills.
	Agriculture	Disruption of natural habitats, soil erosion, fertilizer run-off, poisonous pesticides.
	Forestry	Habitat disruption, soil erosion, pesticides.
Processing	Manufacturing	Soil and underground water pollution, loss of soil productivity, air and water pollution.
Energy	Energy conversion and transmission	Depletion of resources, thermal and radioactive pollution, disruption of land for transmission rights of way. Litter, land use-use disruption, release of hazardous wastes, dust, etc.

The expansion of mining activities in many countries has now become a major cause of land pollution due to loss of soil and destruction of land. Mining has become a threat to the environment because it leads to the huge quantity of waste generally not useful to man and its reprocessing is not economical. Mining operations produce about 1.35 billion tonnes of debris each year.

Urban Wastes

Urban wastes that result in residential areas cause contamination of the soil at places where the wastes are not properly disposed. Wastes like glass, plastic, human excreta, fuel residues, metals and vehicular products are common urban wastes.

Not only does accumulation of these wastes result in poor human health, they also cause pollution of the soil. Urban wastes do not dispose easily, and therefore cause a lot of harm to the soil and its properties. Non bio-degradable wastes like plastic, metal cans and glass cause great harm to the soil.

Effects

i.   Urban wastes dirty residential areas, resulting in the growth of insects and pathogens.

ii. They are harmful to human health.

Agricultural Practices

Agricultural practices also pollute the soil. According to an estimate, agricultural activities produce more than 1.8 billion tonnes of waste, each year.

About three-quarters of this is manure. Much of this manure is piled in dumps where it pollutes streams and waterways. Yet, at the same time, farmers across the continent are suffering from worn-out and depleted soils. Other agricultural wastes include branches and slash left over from logging apart from animal wastes.

In addition to fertilizers and pesticides, soil conditioners and fumigants are used in agriculture. Organic compounds containing lead, mercury and arsenic, when applied to a land, accumulate on the soil permanently and introduce these toxic metals into plant products.

Radioactive Pollutants

Radioactive substances resulting from explosions of nuclear testing laboratories and industries giving rise to nuclear dust radioactive wastes penetrate the soil and accumulate giving rise to land/soil pollution.

Example

1. Radio nuclides of Radium, Thorium, Uranium, isotopes of Potassium (K-40) and Carbon (C-14) are commonly found in soil, rock, water and air.

2. Explosion of hydrogen weapons and cosmic radiations include neutron, proton reactions by which Nitrogen (N-15) produces C-14. This C-14 participates in Carbon metabolism of plants which is then into animals and human beings.

3. Radioactive waste contains several radio nuclides such as Strontium90, Iodine129, Cesium-137 and isotopes of Iron which are most injurious. Strontium gets deposited in bones and tissues instead of calcium.

4. Nuclear reactors produce waste containing Ruthenium-106, Iodine-131, Barium140, Cesium-144 and Lanthanum-140 along with primary nuclides Sr-90 with a half-life 28 years and Cs-137 with a half-life 30 years. Rain water carries Sr-90 and Cs-137 to be deposited on the soil where they are held firmly with the soil particles by electrostatic forces. All the radio nuclides deposited on the soil emit gamma radiations.

5. Biological agents – Soil gets a large amount of human, animal and bird excreta which constitute a major source of land pollution by biological agents.

    Ex: 1. Heavy application of manures and digested sludge can cause serious damage to plants within a few years.

Biological Agents

Biological agents like bacteria, fungi, virus and protozoans are a major cause of soil pollution. Human and animal excreta, poor sanitary conditions, wastes from hospitals and food joints cause soil pollution because they perpetrate growth of biological agents in the soil.

Effects

i.   They cause diseases in human beings.

ii. They harm the development and existence of flora and fauna.

iii. They spoil qualities of fruits and vegetables grown in the polluted soil.

Effect of Soil Pollution

The contamination or degradation of soils impacts heavily on the health of plants. Humans are also affected in numerous ways either directly or indirectly. Polluted soil can harm humans by making contact with the soil or consuming vegetation produce from contaminated soils. Children are even more susceptible to the harms of soil pollution since they spend most of their time playing in close contact with the soil. Thus, soil pollution has a long list of effects.

Endangering Human Health

More than 70% of the soil pollutants are carcinogenic in nature, intensifying the chances of developing cancer in the humans exposed to the polluted soils. Long-term exposure to benzene and polychlorinated biphenyls (PCBs), for instance, is linked to the development of leukemia and liver cancer respectively.

Soil pollutants can also cause skin diseases, muscular blockage, and central nervous system disorders. Humans can be affected indirectly due to bioaccumulation or food poisoning. It happens when people consume crop produce that is grown in the polluted soils or when they consume animal products that eat plants from polluted soils. As a result, humans suffer from acute illnesses and may experience premature death.

For example, high concentrations of lead or mercury in the soil can endanger the functionality of kidneys and liver. It can also hamper brain development in children and cause adverse neurological disorders.

Economic Losses

Crops and plants grown on polluted soils can accumulate poison to an extent deemed unfit for human consumption. Consequently, it leads to enormous economic losses. In some parts of the world, heavily polluted soils with metals and chemicals such as lead, asbestos, and sulfur are considered unfavorable for crop production and cannot be used to grow crops.

The crops grown in the soils and the nearby lands are often poisoned with heavy metals and chemicals thus, discarded after harvesting because of high toxicity levels. According to China's agricultural sector, for instance, about 12 million tons of polluted grains are subjected to disposal on an annual basis, costing Chinese farmers economic losses of up to 2.6 billion U.S. dollars.

Air and Water Contamination

Polluted soil by natural means contributes to air contamination by discharging volatile compounds into the atmosphere. So, the more the toxic contaminants in the soil, the higher the level of toxic particles and foul gasses emitted into the atmosphere. Soil pollution can also lead to water pollution if the toxic chemicals and materials like dangerous heavy metals leach into groundwater or contaminate storm water runoff, which reaches lakes, rivers, streams, or oceans.

Effect on Plant Life

When soils are repeatedly contaminated and accumulate large amounts of poisonous materials and chemicals, the soil reaches a point where it cannot support plant life. Soil pollutants interfere with soil chemistry, biology, and structure. When these changes occur, beneficial soil bacteria, soil microorganisms, soil nutrients, and soil chemical processes begin to deteriorate to an extent where they diminish soil fertility.

The ecological balance is lost completely. On this basis, the soil becomes unsuitable for crop survival or any other form of vegetation. If the plants die, then it means animals dependent on the plants will also die. This leads to migration of the larger animals and predators to other regions to find food supply, gradually leading to a reduction in wildlife and extinction.

Soil pollution can as well change plant metabolism and lower crop productivity. Besides, when plants take up the soil contaminants, they pass them up the food chain, endangering the health of animals and humans.

Acidification

Soil pollution allows emission of relatively large quantities of nitrogen via denitrification, volatilization of ammonia, and the decomposition of organic materials in the soil. As a result, this releases sulfur compounds and sulfur dioxides into the atmosphere, causing acid rain.

In the long-run, it leads to a continuous cycle of pollution whereby the acid rain reduces soil chemistry and nutrients, which would further contribute to ecological balance disturbance and soil erosion. Furthermore, acidic conditions hinder soil ability to cushion changes in the soil PH, leading to the death of plants due to unfavorable conditions.

Diminished Soil Fertility

The most evident and crucial element of the soil is its fertility. Once the soil is contaminated with chemicals and heavy metals or degraded due to human activities such as mining, its fertility depreciates and might even be lost entirely. The harmful chemicals and heavy metals in the soil decrease soil microbial and chemical activity.

The chemical elements can also denature active enzymes that revitalize healthy soil activities. Soil acidification as a consequence of pollution also leaches away essential natural minerals like magnesium and calcium.

Changes in the Soil Structure

Acidification, diminished soil fertility, and death of soil organisms in the soil can lead to changes in soil structure. This is because soil microorganisms help in breaking down organic matter that promotes soil structure regarding vitality and water penetration as well as retention.

Increase in Soil Salinity

The increase in soil salinity, salinization, is an effect of salt accumulation in the soil. Salts occur naturally in the soil. However, increased accumulations are linked to soil pollution. Irrigation and agricultural processes that discharge nitrate and phosphate deposits in the soil are the primary contributors to increasing salt levels in the soil.

Increased soil salinity makes it difficult for plants to absorb soil moisture and reduces groundwater quality. Crops and plants grown in these regions combined with other soil pollutant effects are highly poisonous and can cause severe health disorders when consumed.

Soil Pollution Control

Soil may be polluted and converted into acidic soil or alkaline soil. It should be corrected by suitable technology, before cultivation.

Methods of Soil Treatment

Air sparging is an in situ remedial technology that reduces concentrations of volatile constituents in petroleum products that are adsorbed to soils and dissolved in groundwater. This technology, which is also known as "in situ air stripping" and "in situ volatilization," involves the injection of contaminant-free air into the subsurface saturated zone, enabling a phase transfer of hydrocarbons from a dissolved state to a vapor phase. The air is then vented through the unsaturated zone. Air sparging is most often used together with soil vapor extraction (SVE), but it can also be used with other remedial technologies.

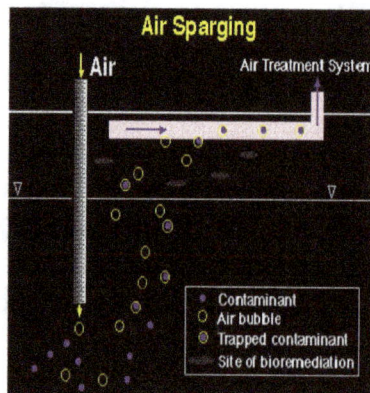

Soil washing is a water-based process for scrubbing soils ex situ to remove contaminants. The process removes contaminants from soils in one of the following two ways:

- By dissolving or suspending them in the wash solution (which can be sustained by chemical manipulation of pH for a period of time); or

- By concentrating them into a smaller volume of soil through particle size separation, gravity separation, and attrition scrubbing (similar to those techniques used in sand and gravel operations).

The concept of reducing soil contamination through the use of particle size separation is based on the finding that most organic and inorganic contaminants tend to bind, either chemically or physically, to clay, silt, and organic soil particles. The silt and clay, in turn, are attached to sand and gravel particles by physical processes, primarily compaction and adhesion. Washing processes that separate the fine (small) clay and silt particles from the coarser sand and gravel soil particles effectively separate and concentrate the contaminants into a smaller volume of soil that can be further treated or disposed of. Gravity separation is effective for removing high or low specific gravity particles such as heavy metal-containing compounds (lead, radium oxide, etc.). Attrition scrubbing removes adherent contaminant films from coarser particles. However, attrition washing can increase the fines in soils processed. The clean, larger fraction can be returned to the site for continued use. Soil washing is generally considered a media transfer technology. The contaminated water generated from soil washing are treated with the technology(s) suitable for the contaminants. The duration of soil washing is typically short- to medium-term.

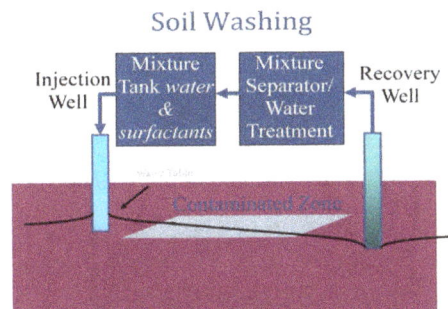

Soil Washing

Biopile treatment is a technology in which excavated soils are mixed with soil amendments and placed on a treatment area that includes leachate collection systems and some form of aeration. It is used to reduce concentrations of petroleum constituents in excavated soils through the use of biodegradation. Moisture, heat, nutrients, oxygen, and pH can be controlled to enhance biodegradation.

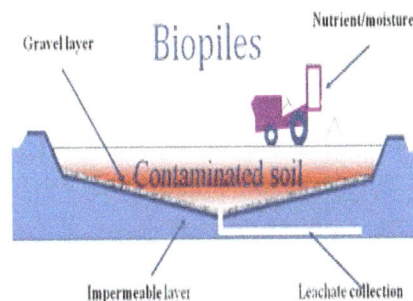

Biopiles

The treatment area will generally be covered or contained with an impermeable liner to minimize the risk of contaminants leaching into uncontaminated soil. The drainage itself may be treated in a bioreactor before recycling. Vendors have developed proprietary nutrient and additive formulations and methods for incorporating the formulation into the soil to stimulate biodegradation. The formulations are usually modified for site-specific conditions.

Soil piles and cells commonly have an air distribution system buried under the soil to pass air through the soil either by vacuum or by positive pressure. The soil piles in this case can be up to 20 feet high (generally not recommended, 2-3 meters maximum). Soil piles may be covered with plastic to control runoff, evaporation, and volatilization and to promote solar heating. If there are VOCs in the soil that will volatilize into the air stream, the air leaving the soil may be treated to remove or destroy the VOCs before they are discharged to the atmosphere. It is a short-term technology. Duration of operation and maintenance may last a few weeks to several months.

Land Farming is a bioremediation treatment process that is performed in the upper soil zone or in bio-treatment cells. Contaminated soils, sediments, or sludges are incorporated into the soil surface and periodically turned over (tilled) to aerate the mixture. This technique has been successfully used for years in the management and disposal of oily sludge and other petroleum refinery wastes. In situ systems have been used to treat near surface soil contamination for hydrocarbons and pesticides. The equipment employed in land farming is typical of that used in agricultural operations. These land farming activities cultivate and enhance microbial degradation of hazardous compounds.

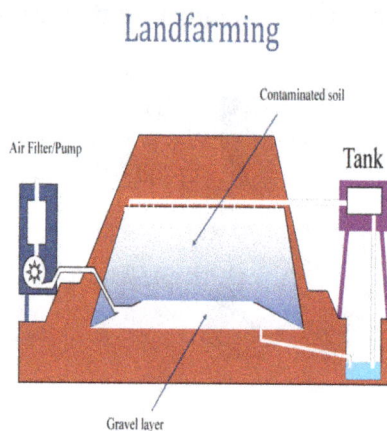

Landfarming

Soil Conservation

Soil conservation is the protection of soil against excessive loss of fertility by natural, chemical, or artificial means. It encompasses all management and land-use methods protecting soil against degradation, focusing on damage by erosion and chemicals. Soil conservation techniques can be achieved through crop selection and rotation, fertilizer and lime application, tilth, residue management, contouring and strip cropping, and mechanical methods (e.g., terracing).

Biological Methods

1. Agronomic practices

- Contour farming
- Mulching

- Crop rotation
- Strip cropping

2. Dry farming

3. Agrostological methods

- Lay farming
- Retiring of land to grass

Mechanical methods

1. Basin listing
2. Contour terracing

Other methods

1. Gully control
2. Afforestation

Terracing – increases the amount of land used for cultivation on steep slope and mountains and reduces erosion.

Soil Amelioration

Amelioration of Acidic Soil: Soil acidity is due to the accumulation of H^+ ions over OH^- ions. Limiting material – are neutralization of H^+ ions such as

- Quicklime- oxide of lime is usually known as burned lime or quicklime.
- Slaked lime-can be obtained by adding water to quick lime.
- Blast furnace slag- a byproduct during the manufacturer of pig iron viz, calcium silicate.
- Basic Slag- is a byproduct of the basic open heart method of producing steel from pig iron,

- Electric furnace slag- is produced from the electric furnace reduction of phosphate rock during preparation of phosphorous. The product is manly the calcium silicate.

The other methods which could result in amelioration of acidic soil are:

- Use of basic fertilizers such as sodium nitrate reduces the soil acidity.

- Proper soil and water management.

- Usage of coral shell, chalk, wood ash, press mud, byproduct material of paper mills, sugar factories, fly ash and sludge etc.

Amelioration of Saline and Alkali Soil

Saline soil- they contain an excess of soluble salts. Saline soil reclamation can be achieved by:

- Providing proper drainage

- Using salt free irrigation water

- Use of acidic fertilizers-such as ammonium sulphate

- Use of organic fertilizers

- Use of organic manures.

Alkaline soil-they contain appreciable amounts of soluble salts. Alkali soil reclamation may be achieved by the following practices:

- Application of gypsum

- Use of sulphur

- Addition of organic matter

- Addition of molasses.

Prevention of Solid Waste Dumping

Open dumping of solid waste should be segregated and recyclable materials could be recycled. Other garbage can be converted into organic manure by suitable technology.

Usage of Bio-fertilizers and Bio-pesticides

Organic/Sustainable Agriculture

Organic farming is a holistic approach which aims for the production of quality and safe agriculture products for consumption. This system requires less financial and external inputs and provides sustainable income to the farming community. Organic farming aims at production of quality and safe agricultural products which contain no chemical residues due to the adoption of eco-friendly production methods and farming systems that restore and maintains soil fertility.

Organic farming is a production method which does not pollute the soil and ground water with chemical residues and provides safe and quality food for consumption. It also increases the biological diversity of plants and animals that helps to maintain the natural eco balance. This approach also aims to recycle only the natural resources and restricts the use of external inputs which indirectly helps to reduce the energy consumption in the farming system considerably.

Concepts of Organic Farming

Organic farming aspires to a complex mix of agronomic, environmental, agricultural and processing and is based on a number of principles. They are:

- To produce food of high quality and safety

- To interact in a constructive and life-enhancing way with natural systems and cycles

- To consider the wider social and ecological impact of the organic production and processing system

- To encourage and enhance biological cycles within the farming systems, involving microorganisms, soil flora and fauna, plants and animals

- To develop a valuable and sustainable aquatic ecosystem

- To maintain and increase the long term fertility of soils

- To promote the healthy use and proper care of water, water resources, and all life therein

- To use, as far as possible, renewable resources in locally organized production systems

- To create a harmonious balance between crop production and animal husbandry

- To minimize all forms of pollution

- To process organic products using renewable resources

- To produce fully biodegradable organic products.

These principles are given equal importance as that of other economically viable production technologies.

Organic Farming Requirements

Achieving the above mentioned principles of organic farming needs a holistic farming system with integrated approach in all aspects. The basic principle of organic farming in enhancing the soil fertility can be achieved through proper recycling of organic wastes, versatile crop rotation and cropping systems, a wide range of biological methods for control of pests, diseases and weeds and to avoid the use of synthetic fertilizers, chemical pesticides and herbicides. Habitat development is the key factor in restoring the natural eco-system which in turn facilitates the symbiotic co-existence of fauna and flora apart from promoting natural predators, parasites etc.

Maintaining Soil Fertility

- Depletion of soil organic matter under intensive cropping system is the key factor in altering biological equilibrium of the soil ecosystem. It is essential to maintain the soil food web, where all the soil organisms viz, bacteria, fungi, actinomycetes, protozoa, earthworms etc., and they flourish in population in the presence of sufficient amount of soil organic matter. In order to maintain the soil fertility, the following farming practices are recommended.

- Increased use of organic manures, green manures

- Enriched vermin-compost and bio composts

- Use of bio fertilizers

- Crop rotation with high and low biomass crops

- Avoiding the use of chemical fertilizers

Plant Protection Methods

Indiscriminate use of chemical pesticides and herbicides leads to soil and ground water contamination which causes health problems in living systems. The accumulation of toxic residues in the food products has created considerable awareness among the producers and consumers. The reports on the pesticides residue in food products revealed that, most of the food products from conventional agriculture contain more than 70 per cent residues. In addition, it also impairs the soil micro flora that is essential to maintain soil fertility. These problems can be solved by adopting organic farming practices which uses only the natural bio pesticides for plant protection. Generally bio pesticides, bio control agents, plant extracts etc are used for controlling the pest and disease problems.

Animal Husbandry

The basis for including animal husbandry in the system is to respect the physiological and ecological needs of animals. This is achieved by providing sufficient quantities of good quality organic fodder, Shelters according to their behavioral needs and also by proper veterinary treatment. Animals are an important part of organic system because they act as the agents for recycling of byproducts with value addition. Further contribute to complete the nutrient cycle and maintaining soil fertility. They also contribute draught energy for agricultural operations and provide essential manure for soil nutrition and urine for pesticides.

Processing of Organic Products

The basis of processing organic products is that as far as possible the vital qualities of the products are maintained throughout each step of the process. This is achieved by choosing and developing methods which are adequate to the specifications of the ingredients and by developing standards which emphasize careful processing methods, limited refining, energy saving technologies, minimal use of additives and processing aids etc. The production and handling of organic products in a safe way can be achieved by adopting existing standards or by developing new standards, which define the safe methods of waste management in the form of products besides packing systems and energy saving systems in processing and transport.

Permissions

All chapters in this book are published with permission under the Creative Commons Attribution Share Alike License or equivalent. Every chapter published in this book has been scrutinized by our experts. Their significance has been extensively debated. The topics covered herein carry significant information for a comprehensive understanding. They may even be implemented as practical applications or may be referred to as a beginning point for further studies.

We would like to thank the editorial team for lending their expertise to make the book truly unique. They have played a crucial role in the development of this book. Without their invaluable contributions this book wouldn't have been possible. They have made vital efforts to compile up to date information on the varied aspects of this subject to make this book a valuable addition to the collection of many professionals and students.

This book was conceptualized with the vision of imparting up-to-date and integrated information in this field. To ensure the same, a matchless editorial board was set up. Every individual on the board went through rigorous rounds of assessment to prove their worth. After which they invested a large part of their time researching and compiling the most relevant data for our readers.

The editorial board has been involved in producing this book since its inception. They have spent rigorous hours researching and exploring the diverse topics which have resulted in the successful publishing of this book. They have passed on their knowledge of decades through this book. To expedite this challenging task, the publisher supported the team at every step. A small team of assistant editors was also appointed to further simplify the editing procedure and attain best results for the readers.

Apart from the editorial board, the designing team has also invested a significant amount of their time in understanding the subject and creating the most relevant covers. They scrutinized every image to scout for the most suitable representation of the subject and create an appropriate cover for the book.

The publishing team has been an ardent support to the editorial, designing and production team. Their endless efforts to recruit the best for this project, has resulted in the accomplishment of this book. They are a veteran in the field of academics and their pool of knowledge is as vast as their experience in printing. Their expertise and guidance has proved useful at every step. Their uncompromising quality standards have made this book an exceptional effort. Their encouragement from time to time has been an inspiration for everyone.

The publisher and the editorial board hope that this book will prove to be a valuable piece of knowledge for students, practitioners and scholars across the globe.

Index

www.ingramcontent.com/pod-product-compliance
Lightning Source LLC
Chambersburg PA
CBHW082018190326
41458CB00010B/3217